Digital Twins for Vessel Life Cycle Service

Houxiang Zhang · Guoyuan Li · Hans Petter Hildre

Digital Twins for Vessel Life Cycle Service

Innovation in Maritime Industry

Springer

Houxiang Zhang
Faculty of Engineering
Department of Ocean Operations and Civil Engineering
Norwegian University of Science and Technology
Aalesund, Norway

Guoyuan Li
Faculty of Engineering
Department of Ocean Operations and Civil Engineering
Norwegian University of Science and Technology
Aalesund, Norway

Hans Petter Hildre
Faculty of Engineering
Department of Ocean Operations and Civil Engineering
Norwegian University of Science and Technology
Aalesund, Norway

ISBN 978-981-97-4767-2 ISBN 978-981-97-4768-9 (eBook)
https://doi.org/10.1007/978-981-97-4768-9

This Springer imprint is published by the registered company Springer Nature Singapore Pte Ltd.
The registered company address is: 152 Beach Road, #21-01/04 Gateway East, Singapore 189721, Singapore

Preface

Digitalization has become a key aspect of making the maritime industries more innovative, efficient, and fit for future operations. One of the most attractive aspects is the concept of digital twins, which refers to a digital replica of physical assets, processes, and systems that can be used as advanced tools for design, operation, and maintenance.

Here we will introduce the development of the digital twin ship, which will be a significant scientific and operational achievement for the maritime industry, making efficient and safe offshore operations possible. The research vessel (RV) Gunnerus from the Norwegian University of Science and Technology (NTNU) is used as a testing platform for our research. It is equipped with the latest technology for a variety of research activities within biology, technology, geology, archaeology, oceanography, and fisheries.

Digital twin enables data exchange safely and easily between different subsystems, modules, and various applications. Thus, the twin ship can provide an integrated view of the ship's various physical and behavioral aspects in different stages, and allow simultaneous optimization of functional performance requirements. In addition, it enables advanced control and optimization, e.g., creating more reliable prediction for flexible objectives (time, output, emissions, fuel consumption), and executing day-ahead and long-term planning for operations. Several related applications are presented in the end to confirm the effectiveness of the digital twin-ship system.

This book will be organized as follows. First, the general introduction to maritime digital twins will be given. Co-simulation mechanism, data collection and transmission, and models and sub-domain models as the fundamental enabling technology digital twin ship will be introduced in detail. After that, related digital twin applications and case studies will be presented to show the effectiveness and efficiency, including the on-board support integration, digital twin in ship docking operation and motion control, prognostics and health management (PHM) providing the first-hand support in daily ship operation, and digital twin of the whole ship could be used as a valuable tool for pre-operation. In the end, the research challenges will be given from authors' viewpoints.

The research work presented in this book is fully supported by the Knowledge-building Project for Industry *Digital Twins for Vessel Life Cycle Service*, under the grant from the Research Council of Norway. The authors thank all support from DNV AS, Kongsberg Maritime AS, Sintef Ålesund, and Ålesund Kunnskapspark AS. Special thanks should be given to the Offshore Simulator Centre for the technical support and cooperation. Many colleagues from the intelligent systems lab at the Department of Ocean Operations and Civil Engineering in NTNU Ålesund campus give great contributions to the research activity and book writing. The thanks should be given to Prof. Vilmar Æsøy, Dr. Yingguang Chu, Dr. André Listou Ellefsen, Dr. Robert Skulstad, Dr. Lars Ivar Hatledal, Dr. Tongtong Wang, Dr. Pierre Major, Dr. Peihua Han, Dr. Baiheng Wu, Dr. Runze Mao, Dr. Chunlin Wang and Ph.D. candidate Finn Tore Holmeset.

Aalesund, Norway

Houxiang Zhang
Guoyuan Li
Hans Petter Hildre

Contents

About the Authors

Houxiang Zhang received the Ph.D. degree in mechanical and electronic engineering from Beihang University in 2003 and the Habilitation in informatics from the University of Hamburg, Germany in 2011. He is currently a full Professor with the Department of Ocean Operations and Civil Engineering, Faculty of Engineering, Norwegian University of Science and Technology (NTNU), Norway. From 2011 to 2016, he also held a Norwegian National GIFT Professorship on product and system design funded by the Norwegian Maritime Centre of Expertise. He has engaged in two main research areas, including control, optimization, and AI application, especially on autonomous vehicle; and marine automation, digitalization, and ship intelligence. He has applied for and coordinated more than 30 projects supported by the Norwegian Research Council (NFR), German Research Council (DFG), EU, and industry. In these areas, he has published over 300 journals and conference papers as the author or co-author. He was nominated and selected as a member of the Norwegian Academy of Technological Sciences in 2019, and a member of the Academy of the Royal Norwegian Society of Sciences and Letters in 2024.

Guoyuan Li received the Ph.D. degree in computer science from the Department of Informatics, Institute of Technical Aspects of Multimodal Systems, University of Hamburg, Hamburg, Germany, in 2013. In 2014, he joined the Department of Ocean Operations and Civil Engineering, Intelligent Systems Laboratory, Norwegian University of Science and Technology (NTNU), Ålesund, Norway, where he is currently a Professor in ship intelligence. He has published more than 150 articles in the areas of his research interests, including the modeling and simulation of ship motion, autonomous navigation, intelligent control, optimization algorithms, and the locomotion control of bioinspired robots.

Hans Petter Hildre is a Professor and Head of the Department of Ocean Operations and Civil Engineering at the Norwegian University of Science and Technology (NTNU). His area of interest is product design and system architecture design. He is the Director for the Centre for Research Driven Innovation (SFI-MOVE) within marine operations. This is cooperation between NTNU, SINTEF, University Sao

Paulo, and 15 companies at the west coast of Norway. He is the Head of Research in the national program Global Centre of Expertise Blue Maritime, a Project Leader in several research projects, a Member of the Board of five companies, and has a number of patents.

List of Figures

List of Tables

Chapter 1
Introduction

Houxiang Zhang, Guoyuan Li, and Hans Petter Hildre

1.1 Norwegian Maritime Industry

Norwegian maritime industrial cluster is a world leader in developing complex, customized ships and offshore vessels for the global market, particularly for demanding operations, where safety and environment regulations are focused areas [1]. Norwegian ship-owners control the 6th largest fleet in the world, and the international shipping fleet consists of about 1700 vessels. Norway's share of the world fleet in terms of value is about 6%. Compared to the fact that Norway only has 0.1% of the world's population it's obvious that the maritime sector is very important to Norway. Actually the maritime sector accounted for 20% of Norway's total exports and 40% if we exclude oil and gas. This makes the maritime sector the second most important export sector and amounts to 230 billion NOK.

The Norwegian government has produced a report "Maritim21" with the subtitle "*An integrated maritime strategy for research, development and innovation*" [2]. The report states: "*Greater investment in research, development and innovation for the maritime sector is needed to uphold Norway's status as a leading maritime nation." and "In the years ahead it will be critical to implement measures to ensure that the Norwegian maritime industry and research and education environments develop specialist expertise in enabling technologies and are equipped with the knowledge and innovation experience to apply these*". The maritime industry in Norway has a complete value chain of expertise and activities, including design, building, equipping and operation of advanced offshore vessels. Companies stakeholders such as Kongsberg Maritime, DNV, Vard AS have world-class training simulation centers for marine operations. Leading know-how on systems and operational modelling/simulation are of vital strategic importance for this industrial cluster in order to further strengthen their global competitiveness. Maintaining technology leadership and international competitiveness is imperative for securing future employment in the national maritime industry and other sectors supplying the industry with know-how, services and equipment.

H. Zhang et al., *Digital Twins for Vessel Life Cycle Service*,
https://doi.org/10.1007/978-981-97-4768-9_1

Today's maritime engineering systems are operating in highly dynamic environment. The challenge is to develop a concept leveraging on the different levels of system specific services already provided by manufacturers, where safety and efficient performance of complex integrated systems can be managed from the early stages of a new vessel manufacturing project and throughout the vessel's life cycle. In order to be able to respond fast to unexpected events and anomalies, future systems need to become more autonomous. Advanced marine systems should be able to decide between different actions, adapt to dynamic environments and execute high-level task specifications without explicitly being programmed. To meet these requirements, control system will need access to very realistic models of the current state of the process and their own behavior in interaction with the real-world environment. From a simulation point of view, digital twins are the next wave in modeling, simulation and optimization technology. It will be a significant scientific and operational achievement for the maritime industry, keeping Norway on the technological lead.

Modern marine vessels are increasingly operated autonomously through strongly interacting subsystems [3]. Systems and sub-systems exchange data and make coordinated operational decisions, ideally without any user interaction. Designing, operating and life cycle service to support such vessels is a complex and intricate engineering task requiring an efficient development approach to consider the mutual interaction between subsystems and the inherent multi-disciplinarity. Scalable simulation technologies should take the lead in this process. Furthermore, the work flow in maritime industry does not stop after vessel delivery. Through system updating or due to life cycle maintenance, subsystems can be changed and updated. To make sure of that, product design and useability need to be coupled already during early stages of design, which requires traceability through a performance data management system that spans the entire vessel life cycle.

The recent years have seen an increasing interest in developing and employing digital twins [4], big data [5] and cloud computing for maritime industrial system design, ship intelligence, and operational service. The report prioritises the importance of enabling technologies in order to perform a digital transformation of the maritime industry. Digitalization has become a key aspect of making the maritime industries more innovative, efficient and fit for future operations. Increased use of advanced tools for designing and evaluating system performance, safety and structural integrity are generating a range of digital models of a vessel and its equipment. In the operational phase, cheaper sensors and increased connectivity together with increasing data storage and computational power enable for new ways of managing a vessel's safety and performance. Digital twins for marine product design and operational autonomy are given great emphasis and can be exemplified with this quote: "*Autonomy, automation and remote control are very promising in terms of cutting costs and enhancing the safety of operations, and may make sea transport competitive in entirely new market segments* [2]."

1.2 Digital Twin in Life Cycle Service

A digital twin serves as a virtual replica designed to mirror a physical product, capturing specific aspects or behaviors. It's important to note that building a single comprehensive digital twin to encompass all facets of a physical artifact is a complex task. Sometimes it is even impossible. Digital twins not only facilitate product design but also play a significant role in the entire product life cycle service.

During the design phase of complex products like ships, engineers create numerous models to address various aspects. For instance, one such model pertains to the fatigue lifetime of components, exemplified by the gearbox of a crane on an offshore wind service vessel. In this process, designers define models, assume load conditions, and conduct fatigue analyses to estimate the component's lifetime. Subsequently, based on the estimated fatigue lifetime, a service interval for replacing parts of the system is determined. The conventional approach utilizes real loading data from the gearbox to continuously update fatigue life assessments and corresponding service intervals. The actual loading conditions depend on various factors specific to each vessel, such as the lifting payload, frequency of use, and prevailing weather conditions during crane operation.

Developing the digital twin concurrently with the original models offers significant time savings. Figure 1.1 introduces a theoretical framework for product design and the corresponding life cycle. This process begins with identifying a business opportunity based on market needs, followed by development phases, including conceptual design, engineering design, and production. The successful development of complex systems like vessels necessitates close collaboration among design firms, equipment suppliers, shipyards, and ship owners. A deep understanding of the product's life cycle, particularly its usage scenarios, is crucial for effective ship design. Insight into life phases and usage situations is imperative to ensure optimal design decisions throughout the development process.

A service-oriented development process necessitates consideration of two interconnected life cycles: the product life cycle and the customer activity cycle. The product life cycle encompasses the various phases a product undergoes throughout its lifespan. In parallel, the customer activity cycle outlines the recurring user

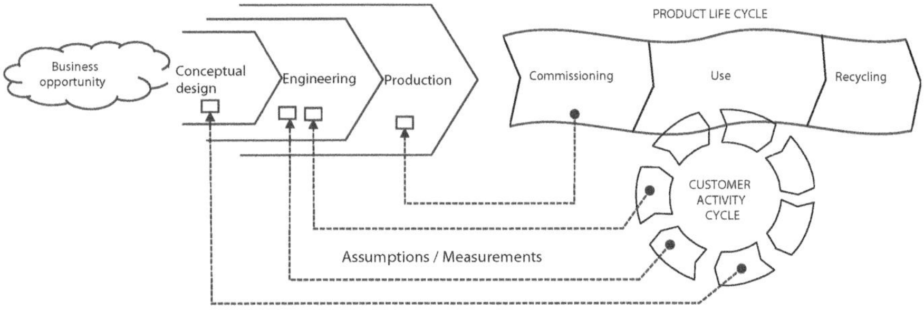

Fig. 1.1 Development process and following product life cycle phases

operations associated with the product. The vessel's performance is contingent upon the effectiveness of its design in facilitating quality experiences within the customer activity cycle, which typically comprises repeated user activities. Service provision is integral to supporting the product's business objectives and accommodating customer actions. This paradigm offers designers an opportunity to augment the product's utility and value through service enhancements. Designers create models that represent performance within the customer activity cycle, as depicted by feedback loops.

Let's delve deeper into an example and analyze the customer activity cycles for the wind service vessel. The service activity cycle outlines the essential tasks the vessel must undertake, as depicted in Fig. 1.2. Critical aspects pertaining to the life phases and utilization of the products are meticulously incorporated into the design and engineering processes. Models are formulated based on assumptions regarding the operational conditions of the product. The left part of Fig. 1.2 depicts the service activity cycle for the crane gearbox, while the right part illustrates the utilization of the gangway for transferring repair crews to offshore wind turbines.

The transition from a predictive model based on assumptions to a digital twin is depicted in Fig. 1.3. This concept can be elucidated through the example of the crane gearbox service cycle. By measuring the load on the gearbox, a cumulative fatigue calculation can be performed using actual loading conditions. However, in certain situations where the required characteristics cannot be directly measured, supplementary calculations are necessary.

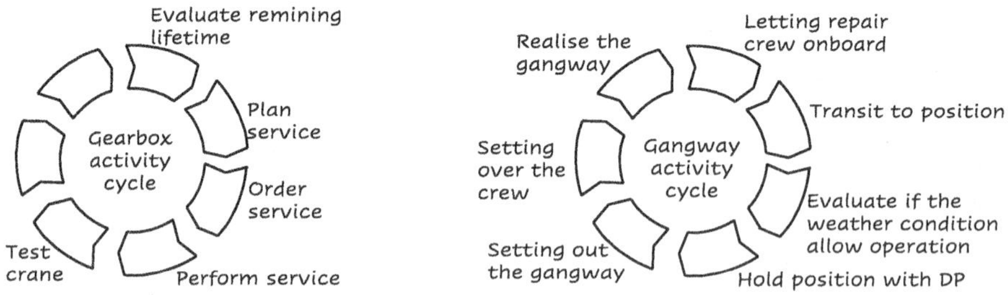

Fig. 1.2 Customer activity cycle (Taking crane gearbox and gangway as examples)

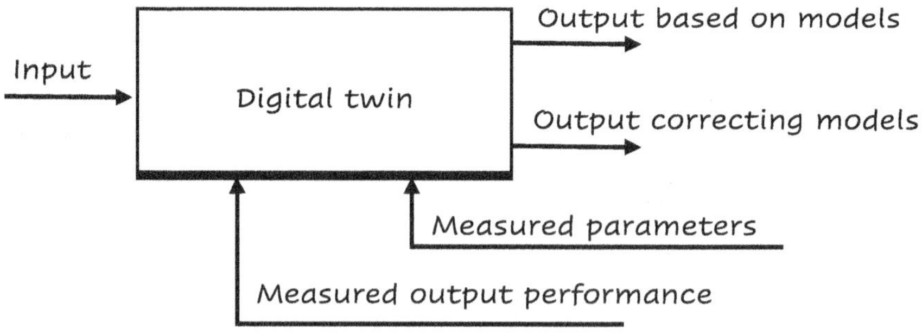

Fig. 1.3 From model based approach to digital twin

Models do not always yield accurate results, even when parameters are measured. In some cases, it may not be feasible to measure the desired behavior directly, leading to measurements of alternative behaviors. However, data-driven methods can help mitigate inaccuracies in models. For example, consider the gangway activity cycle depicted in the right part of Fig. 1.2. The ship's ability to maintain its position using dynamic positioning (DP) relies on various characteristics that must be assumed or verified through testing. These include thruster dynamics, wind, wave, and current, which are all behaviors that cannot be accurately evaluated. Additionally, the vessel's unique loading condition presents challenges for prediction. Furthermore, limitations on deploying the gangway are influenced by factors such as the vessel's roll, which can cause accelerations at the gangway's tip. However, the roll is affected by its movements due to incoming waves, which are challenging to predict accurately due to inaccuracies in vessel hydrodynamic models. To address these challenges, a hybrid modeling technique can be employed to compensate for inaccuracies in the model, as illustrated in Fig. 1.3.

1.3 Research Vessel Gunnerus

A digital twin of the research vessel (R/V) Gunnerus will be presented in this book [6]. Here, we will first give an introduction to the platform. Gunnerus as seen in Fig. 1.4, which is owned and operated by NTNU, was put into operation in 2006. It is equipped with the latest technology for a variety of research activities within biology, technology, geology, archaeology, oceanography, and fisheries research [7]. In addition

Fig. 1.4 Research vessel (R/V) Gunnerus. © 2024 Fredrik Skoglund, reprinted with permission

Table 1.1 Physical parameters of the R/V Gunnerus

Description	Parameters	Short version	Elongated version
Length over all	L_{oa} $[m]$	31.25	36.25
Length between perpendiculars	L_{pp} $[m]$	28.9	33.9
Breadth middle	B $[m]$	9.6	9.6
Draught middle	T $[m]$	2.7	2.7
Design speed	U $[knot]$	9.6	9.6

to research, the ship is used for educational purposes and is an important platform for marine courses at all levels and disciplines. The ship is fitted with a dynamic positioning system and a HiPap 500 unit. This provides optimal conditions for ROV operations and the positioning of any deployed equipment. The diesel electric system of Gunnerus has been specially designed for low hydroacoustic noise levels to accommodate testing and development of hydroacoustic equipment. The vessel is arranged with wet lab, dry lab and a computer lab in addition to a large aft deck. Accommodation comprises three double berth scientific personnel cabins and three single berth crew cabins. The large mess hall functions as a lecture room for 25 people.

Table 1.1 holds the main dimensions of the vessel. The R/V Gunnerus was equipped with twin fixed-pitch ducted propellers and rudders and one tunnel thruster from Brunvoll. The diesel electric system of Gunnerus has been specially designed for low hydroacoustic noise levels to accommodate testing and development of hydroacoustic equipment. In 2015, the R/V Gunnerus went through a thruster refit, and the original propellers were replaced with Permanent Magnet (PM) rim-drive azimuthing thrusters, and its layout is presented as Fig. 1.5. The original propellers were 5-bladed, high skew type with a diameter of 2.0 m that rotated in a 19A type duct profile, and the new azimuthing thrusters incorporates a ring propeller in a tailor-made duct with a diameter of 1.9 m with four blades having a forward skewed shape. Figure 1.6 shows

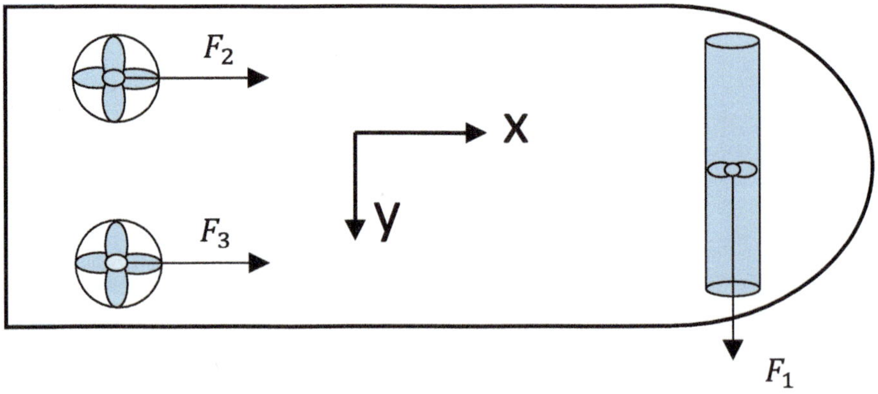

Fig. 1.5 The thruster configuration the R/V Gunnerus

Fig. 1.6 The propulsion arrangement of the R/V Gunnerus before and after retrofit

the propulsion configuration on Gunnerus before and after retrofit, where the left is the origin pitch propeller with ice-fins, and the right is the refitted azimuth thruster provided by Rolls-Royce. The same diesel-electric system supplied the propulsion and maneuvering power before and after the conversion. In 2019, the vessel was elongated to 36.25 m from original 31.25 m. All other geometrical and propulsion characteristics except the length dimensions remained the same. Because of the elongation, there are two versions of the vessel, which are both used in this project.

As a research vessel, there are many data acquisition systems installed onboard the Gunnerus, like thrusters, power generators, navigational sensors, and motion control systems, etc. (see Chap. 4 for details).

1.4 Digtial Twin Gunnerus Ship

The digital ship can be either deployed on its physical counterpart, or placed onshore in a remote control center given the sensor data is able to be transmitted via gateway. A historical dataset can be generated by taking all-year measurements from the onboard sensors placed on the physical ship, e.g., the ship hull, the engine and the thruster. Digital models of systems and sub-systems in simulation can thus be refined using combined model-driven and data-driven approaches [8] to ensure the fidelity of the digital twin of a ship. Based on the digital ship simulation, a variant of control levels, from remote control to fully autonomy can be implemented in the digital twin system. As a result, different onboard support tools, such as visualization, task configuration and high-level applications like path planning can be included in the system. The result of the onboard support tools will either be fed into simulation for validating operations or be treated as metrics for decision making on the physical ship.

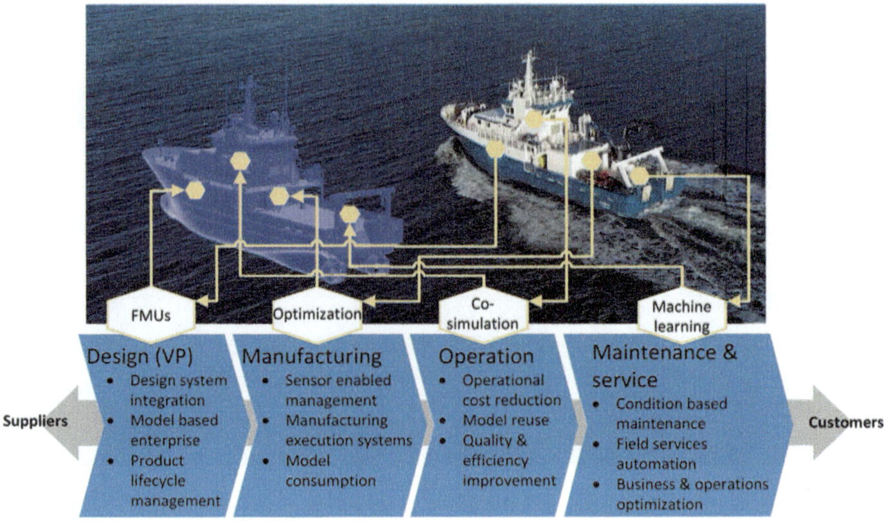

Fig. 1.7 Digital twin of Gunnerus ship

From 2018, our team has been working on establishing a digital twin of the R/V Gunnerus. As illustrated in Fig. 1.7, it is a digital open marine industrial infrastructure not only for overall system design, allowing configuration of systems and verification of operational performance, but also more focusing on providing early warning, life cycle service support, and system behavior prediction. The digital twin could be used in different vessel life cycle service.

- **Design phase**: The life of a ship starts from its design. In this phase, the ship could be modelled in a whole piece including all necessary important systems, so that the designer could find out if the design meets all technical requirements. The hull is designed and such as thrusters, machineries, cranes, winches, automation systems, control systems from many suppliers are put together [9, 10]. The maritime digital twin models should be constructed prior to or in parallel with the actual building of a vessel, enabling virtual prototyping (VP), virtual manufacturing and digital commissioning, integration testing and analysis in the earlier stage. Here, in the digital twin ship, we will only need good enough as long as with trustable fidelity. On the contrary, designers could easily find any unexpected errors or mistakes in this very early stage so to revise, modify or polish the system. We will employ a novel co-simulation mechanism as a platform for digital twins. Furthermore, Functional Mock-up Units (FMUs) will be used for component model coupling.
- **Manufacturing phase**: Now the design will come down to the detailed modelling, simulation, and development. Ship system models are created, and several performance indicators are simulated and optimized to fulfil the ship owners need. Building system models from a variety of equipment and vendors is challenging. All models and sub-models should be with high fidelity in order to make

sure the single system with prefect performance. This phase should guarantee all components, systems running as expected. Depending on the complexity level, different models implementations contain different sets of properties describing the behavior of the physical system. Usually, a more complex model implementation provides more realistic results close to the physical system, but it also requires more parameters and state variables for computation. The Functional Mock-up Interface (FMI) standard has become an enabler for integration of simulation models from different disciplines and suppliers.
- **Operation phase**: During the operational and testing phase, the digital twin becomes a system for integration, processing and analysis of the operational data. In this phase, many operational models in digital twin ships could be light-weight and real-time so that ideally, the digital twin will provide decision support and predictions for ship owners, subsystem manufacturers and safety training/improvement for crew and operators [8, 11–13]. It will enable the use of model-based/data-based simulation in combination with sensor measurements from the real vessel, without the need for physical inspections. Here, we should agree that the requirement to real-time on operations imposes an intractable challenge to the microsteps of the simulators, which decide the simulation accuracy and efficiency.
- **Maintenance and service phase**: In the above way, the digital twin tracks information on all parameters to define how each individual module and sub modules behaves over its entire useful life, including the initial design and further refinement, manufacturing related deviations, modifications, uncertainties, updates as well as sensor data from on-board systems, maintenance history and all available historical and fleet data obtained using data mining. All outcomes will be further used for ships that are either autonomous or remote-controlled for safety and reliability enhancement [6]. The digital twin ship could provide a unified solution to save different types of data of critical machinery in a database, further to establish condition based maintenance system [14], control optimization [15], and digital commissioning.

The marine digital twin tracks information on all parameters to define how each individual module and sub modules behaves over its entire useful life, including the initial design and further refinement, manufacturing related deviations, modifications, uncertainties, updates as well as sensor data from on-board systems, maintenance history and all available historical and fleet data obtained using data mining. Using digital twins for marine engineering could bring the following advantages.

- **Reduce development time and enhance production efficiency**: provide an integrated view on the vessel's various physical and behavioural aspects in all stages, allow simultaneous optimization of all functional performance requirements throughout the entire development cycle, from early concept to detailed engineering and commissioning; reduce design tolerances, manufacturing uncertainties and stochastic variabilities of vessel operation.
- **Improve operational flexibility and reduce cost**: create more predictable, faster starts for flexible objectives, execute day-ahead and long-term planning for improved operations, reuse the sub system models, data and operational sources.

- **Enhance the life cycle value chain and improve the system performance and health conditions**: allow data exchange between different sub-systems, modules, and proprietary applications; make remote access systems assure system availability via digital twins, and early warnings utilized for short time technical support.
- **Improve quality and efficiency of maritime product and operation approval and certification processes**: combine simulation models and sensor data on an open platform; facilitate the design and verification of cyber-physical systems; allow classification societies to replace periodical surveys with condition and event based inspections; establish a standardized service for the maritime industry for efficient design and approval process.

1.5 Summary

The digital twin of Gunnerus is to bring together relevant research partners from the academia and industry in an effort to pursue a vision for an open digital twins ecosystem, identifying relevant research questions and providing reliable technological solutions. Due to the nature of this research, expertise and methodologies from complementary disciplines such as systems design, cybernetics, artificial intelligence, computer science and the maritime industry must be combined.

Although the implementation of the digital twin has some current limitations, such as insufficient possibilities for synchronization between the physical and the digital world to establish closed loops, the lack of high fidelity models for simulation and virtual testing at multiple scales [15, 16], the lacking uncertainty quantification for such models, as well as the challenges for gathering and processing large data sets, Fig. 1.7 shows a sound conceptual framework, from which a digital ship can be refined via historical/ onboard sensor data using data-driven methods and be combined with other models used in marine operations for onboard support or advanced control. As a sequence of evolution from simulation, virtual prototyping to digital twin implementation, a twin ship with high fidelity can thus be realized and used as digital tools for either decision making in remote control center, or control basis for autonomous ship. Four key issues in Gunnerus digital twin will be introduced in the following chapters separately, including:

- **Co-simulation mechanism as digital twins platform**;
- **Data collection and transmission**;
- **Models and sub-domain models for co-simulation**;
- **Enabling tools for onboard support on the ship or onshore**.

Houxiang Zhang received the Ph.D. degree in mechanical and electronic engineering from Beihang University in 2003 and the Habilitation in informatics from the University of Hamburg, Germany in 2011. He is currently a full Professor with the Department of Ocean Operations and Civil Engineering, Faculty of Engineering, Norwegian University of Science and Technology (NTNU), Norway. From 2011 to 2016, he also hold a Norwegian National GIFT Professorship on product and system design funded by the Norwegian Maritime Centre of Expertise. He has engaged into two main research areas, including control, optimization and AI application especially on autonomous vehicle; and marine automation, digitalization and ship intelligence. He has applied for and coordinated more than 30 projects supported by Norwegian Research Council (NFR), German Research Council (DFG), EU, and industry. In these areas, he has published over 300 journals and conference papers as the author or coauthor. He was nominated and selected as a member of the Norwegian Academy of Technological Sciences in 2019, and a member of Academy of the Royal Norwegian Society of Sciences and Letters in 2024.

Guoyuan Li received the Ph.D. degree in computer science from the Department of Informatics, Institute of Technical Aspects of Multimodal Systems, University of Hamburg, Hamburg, Germany, in 2013. In 2014, he joined the Department of Ocean Operations and Civil Engineering, Intelligent Systems Laboratory, Norwegian University of Science and Technology (NTNU), Ålesund, Norway, where he is currently a Professor in ship intelligence. He has published more than 150 articles in the areas of his research interests, including the modeling and simulation of ship motion, autonomous navigation, intelligent control, optimization algorithms, and the locomotion control of bioinspired robots.

Hans Petter Hildre is professor and head of the Department of Ocean Operations and Civil Engineering at the Norwegian University of Science and Technology (NTNU). His area of interest is product design and system architecture design. Hans Petter is the director for Centre for Research Driven Innovation (SFI-MOVE) within marine operations. This is cooperation between NTNU, SINTEF, University Sao Paulo and 15 companies at the west coast of Norway. Professor Hildre is head of research in national program Global Centre of Expertise Blue Maritime, project leaders in several research projects, member of the board in 5 companies, and has a number of patents.

Chapter 2
Digital Twins in the Maritime Industry

Houxiang Zhang, Runze Mao, Pierre Major, Guoyuan Li, and Hans Petter Hildre

Digital twin development involves creating a digital replica of a physical system, often employed in multi-domain operation systems (MDOS) to simulate and analyze the system's behavior. By generating a digital twin, a deeper understanding of the system's dynamics can be achieved, facilitating optimization. This technique has become increasingly crucial for optimizing dynamic operation systems.

A digital twin is a virtual representation of a physical asset, enabled through data and simulators, aimed at real-time prediction, optimization, monitoring, control, and improved decision-making [17]. It leverages various digital information sources associated with the asset, including system and data information models, 3D models, mathematical models, dependability models, condition and performance indicators, and data analysis [18]. This comprehensive integration allows for the evaluation of system behavior under diverse operating conditions, facilitating the identification of potential issues and optimization of system performance. There is a pronounced demand for innovation and efficiency across operations, life cycle services, and the design of marine systems. With modern marine vessels increasingly operating autonomously via strongly interacting subsystems, the maritime industry stands to benefit significantly from the adoption of digital twin technology [19].

In essence, digital twins comprise three fundamental components: the physical entity, its corresponding digital representation, and the interaction between these two elements. In our research, the physical entity under consideration is the surface vessel Gunnerus. This encompasses various operational environments experienced by the vessel, including waves, winds, and other pertinent conditions. Gunnerus has undergone extensive development and has been operational for several years.

The second component entails the digital representation of the Gunnerus ship, which involves the comprehensive integration of models and sub-models representing its diverse systems. The quality of models within a digital twin is of utmost importance. However, despite having all requisite models prepared, seamless integration into a cohesive system is not guaranteed. It's worth noting that many ship models are developed using distinct professional or commercial software platforms, lacking

H. Zhang et al., *Digital Twins for Vessel Life Cycle Service*,
https://doi.org/10.1007/978-981-97-4768-9_2

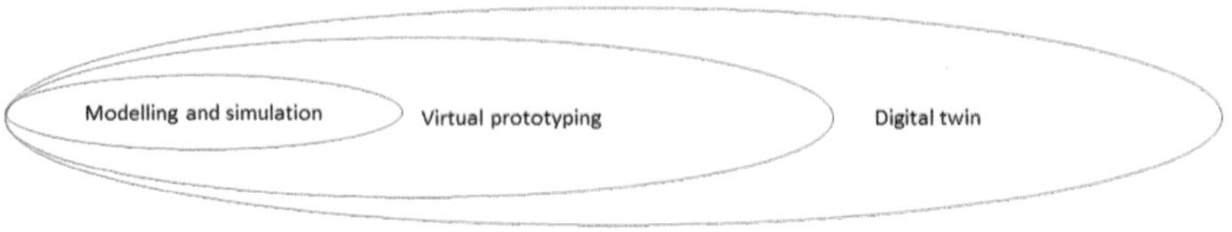

Fig. 2.1 Difference among modelling and simulation, virtual prototyping and digital twins

a standardized interface for communication. Consequently, the integration of such models becomes exceedingly challenging. Thus, the development of a framework for facilitating interfacing becomes imperative. Co-simulation emerges as a promising solution for integrating models and sub-models seamlessly. Upon successful implementation, the amalgamation of these models constitutes what is known as a virtual prototype (VP) or digital shadow (DS) [20]. However, it falls short of being classified as a digital twin.

Completing the digital twin requires establishing communication and interaction between the physical ship and its virtual representation. This involves the transmission of physical sensor data, operational states, and environmental conditions from the physical vessel to its digital counterpart. Concurrently, operational support and planning data are relayed back to the ship to enhance safety and optimize operations. All data exchanges are stored in the cloud for future reference. Figure 2.1 illustrates the distinctions among modeling and simulation, virtual prototype (VP) or digital shadow (DS). Subsequent sections will delve into the various technologies associated with digital twins.

2.1 Modelling and Simulation

Model development for multi-domain operation systems (MDOS) requires a combination of modeling techniques to accurately represent physical systems. Models and sub-models that capture the behavior of physical systems serve as the foundation towards achieving a fully functional digital twin. These digital representations can encompass physics-based models, data-driven models, algorithms, or purely graphical models for visualization purposes.

The last decades have seen an increasing interest in developing computer-based design and analytical tools for different applications. Initially, several general-purpose simulation environments have gained prominence in both research and education. These include MATLAB/Simulink [21], SimulationX [22], Modelica [23], 20-sim [24], COMSOL [25] and Algoryx [26], etc. Additionally, specialized analysis software tailored to specific domains such as structures, hydrodynamics, computational fluid dynamics, power systems, and control systems plays a crucial role in the design process to evaluate the performance of individual subsystems. This category of software is commonly referred to as domain-specific software. Examples include PSCAD for power systems [27], Adams for dynamics [28], OrcaFlex and ABAQUS for structures [29, 30], dSPACE for control systems [31], and GT-Suite for engine systems [32]. Many of these tools offer interoperability through interfaces to facilitate seamless integration. However, employing these tools for system simulation in design and operation requires specialized expertise to effectively leverage their analysis capabilities.

Domain-specific simulation software tools are integral to the design process, particularly for complex dynamic systems, necessitating specialized expertise to leverage their analytical capabilities effectively. These software packages are predominantly standalone, offering limited capacity for system modeling and simulation coupling. The aviation and defense sectors have pioneered advanced methodologies in both design and operation, driven by stringent customer and regulatory mandates. Notably, simulation serves as a dual-purpose tool for operator training and design, leveraging distributed models and simulator sites. The automotive industry, exemplified by initiatives like the European Modelisar project, has demonstrated innovation through the establishment of Modelica and the Functional Mock-up Interface (FMI) standard [33]. While the maritime industry has partially embraced these advancements, primarily in crew training using simulators, the utilization of system simulators for design remains constrained. This limitation is underscored by the unique configurations of individual ships, unlike the standardized builds seen in aircraft and automotive manufacturing.

The approach proposed for modeling in this book is based on Object-oriented Modeling (OOM) and Functional Mock-up Interface (FMI) co-simulation. The primary advantage of this approach lies in the modularization of complex dynamic systems for modeling and efficient coupling of domain-specific simulation tools. This facilitates the reuse of models with variations in complexity and enables interfacing between different simulation tools.

OOM traces its origins to software model development and led to the establishment of the Unified Modeling Language (UML) in the 1990s [34]. Subsequently, Systems Modeling Language (SysML), as an extension of UML, emerged as a graphical modeling language supporting the specification, analysis, design, and verification of dynamic systems [35]. SysML encompasses four pillars: requirements, behavior, structure, and parametric, utilizing nine diagrams to depict various aspects of a system model. Notably, the INTO-CPS project has developed INTO-SysML models for co-simulation of cyber-physical systems (CPS) [36]. The INTO-SysML profile customizes SysML for architectural modeling in a multi-modeling and FMI co-simulation setting, introducing specializations of SysML blocks to represent different CPS components. This hierarchical description of the CPS architecture constitutes the building blocks enabling comprehensive modeling. SysML is a powerful, but formidable, technical language to master, especially for the designers and system engineers [37]. Its extensions, such as INTO-SysML, could be a better alternative for modeling if matured. Currently, the application of OOM within dynamic operation systems is still in its nascent stage. Introducing the OOM method for dynamic operation systems aims to facilitate modularization and integration of model development, enabling diverse users to contribute their expertise on subsystems without delving into the intricacies of others.

More recently, the Modelica language has garnered attention from both academic and industrial users. Modelica is a non-proprietary, object-oriented, equation-based language designed for the convenient modeling of complex physical systems. Basic models of physical components are defined in a declarative manner by their constitutive equations and connected with the outside world without implied causality. This approach makes the description of physical systems more flexible and understandable compared to using causal or block-oriented modeling languages. Additionally, it is easier to use than directly writing simulation code using procedural languages like C or FORTRAN. Complex models can be constructed by connecting these basic models wherever physical connections exist. Since the models are formulated in terms of generic differential-algebraic equations, components defining interactions between different physical domains can be combined without any restriction.

To facilitate the orchestration of a network of models, an extension of the FMI standard has been developed. The System Structure and Parameterization (SSP) of Components for Virtual System Design defines a standardized way to store and apply parameters to these components, as well as a standardized format for the connection structure of a network of components.

The Open Simulation Platform (OSP) represents an open-source industry initiative for co-simulation of maritime equipment, systems, and entire ships. The OSP interface specification (OSP-IS) is an addition to the FMI 2.0 standard for co-simulation, providing a method for adding semantic meaning to model interface variables. OSP-IS facilitates the faster construction of co-simulation simulators by simplifying the model connection process and enabling the validation of semantically correct simulations.

In summary, the models representing physical systems and behaviors must meet simulation requirements in terms of complexity, modularity, flexibility, scalability, and reusability. However, the level of model implementation depends on the specific application to ensure both adequate simulation accuracy and efficiency. For instance, as illustrated in Fig. 2.1, during the ship's preliminary design phase, relatively low-fidelity models suffice for rapid prototyping to conduct quick whole-ship proof-of-concept testing. Subsequently, high-fidelity models are required for detailed design, control, and prediction. Real-time models become necessary during the training phase. The reusability of models across different stages of a system's life cycle is also highly beneficial and valuable. For systems comprising multiple physical domains, a balance must be struck between the number of defined domain models and the scale of the entire system to ensure the flexibility and speed of the digital twin system. For example, the Gunnerus ship's hull is modeled and simulated in Vesim [38]; the power plant is modeled in 20-sim [39]; various deck machineries are modeled in Simulink, simulationX, and 20-sim, respectively [40, 41]; environmental components are coded in Simulink, and operational sensor states are coded in C++; while different controllers are developed in Java or Python. The modeling approach for implementing the Gunnerus twin will be described in Chap. 5.

2.2 Virtual Prototyping

Maritime systems typically communicate using various formats and protocols. The diversity in protocols and the array of installed instruments and systems present challenges when integrating new systems, particularly those requiring data from multiple subsystems. Moreover, configuring each protocol requires explicit knowledge of the location and method for accessing each piece of information. While integrating all models and sub-models into a cohesive whole appears logical, it poses significant challenges in practice.

Two fundamental concepts are introduced initially. Virtual prototyping (VP) is an iterative design method and process that involves stakeholders collaborating around a virtual rendition of a product or service to envision and analyze different aspects and metrics throughout its life cycle [42, 43]. VP for advanced maritime systems and offshore operations entails developing a high-fidelity visual and physics-based representation of the systems and operations, enabling stakeholders to interact, share knowledge, uncover insights, refine procedures, and mitigate operational risks. Virtual prototypes (VPs) are integrations of digital or mathematical models used to simulate the behavior of existing or conceptual systems. VPs are computer simulations of physical products that can be visualized, analyzed, and tested across various stages of the product life cycle, including design/engineering, manufacturing, service, and recycling, as if they were real physical models [44]. In the same vein, VP is akin to the concept of digital shadow (DS) [20] as depicted in the literature.

Advancements in 3D modeling and simulation technologies have markedly enhanced the accuracy and realism of VPs. Consequently, simulating the behavior and performance of prototypes in a virtual environment has become more feasible, accounting for various design and operational parameters. Leveraging VPs enables the early identification and resolution of design and operational challenges during the development phase, thereby diminishing the necessity for physical prototypes and expediting time-to-market. Furthermore, a notable progression in VP involves the fusion of virtual and physical testing methodologies. This integrated approach facilitates the verification and validation of VPs through physical testing, furnishing invaluable insights for refining the virtual model. Consequently, hybrid testing methods have emerged, amalgamating virtual and physical testing to furnish a more comprehensive assessment of prototypes.

The adoption of VPs has been further catalyzed by the advent of cloud-based simulation platforms. These platforms furnish access to high-performance computing resources, enabling the simulation of intricate and large-scale prototypes. Consequently, virtual prototyping has become more accessible and cost-effective, democratizing these technologies even for smaller organizations. Additionally, the integration of VPs with augmented reality (AR) and virtual reality (VR) technologies has substantially enriched the user experience. This integration enables the visualization and interaction with prototypes in a more immersive and intuitive manner, yielding valuable insights into their design and functionality.

Recent research in VP has garnered considerable interest [45]. In one study, VP was employed for feasibility studies aimed at optimizing the towing configuration of a novel, complex-shaped design for a floating wind farm anchor during underwater towing. This research enabled technicians and staff to explore numerous operational details [46]. Another study described a mathematical model for subsea installation based on VR, capable of running in near real-time. However, the wave model in this work only comprised a single wave component, with wind effects disregarded in the VR environment [47]. Another research project, also leveraging VR technology [48], presented a full mission simulator using a similar system based on the physics engine Vortex and the visualization tool Vega Prime. The simulation results were validated against benchmarks from SESAM, a software suite for hydrodynamic and structural analysis of ships and offshore structures, across various wave heights. However, the study did not specify the wave spectrum, but it did account for wind and current effects.

An integrated model combining engine thermodynamic modeling and control system functional modeling was proposed for the development of high-fidelity digital twins of marine engines[49]. The results demonstrate that the developed model can sufficiently represent the engine and its subsystems/components behavior and effectively capture the engine control functionality. However, it involves one-way communication from the physical system to the digital one, leaving the system one step shy of real digital twins. Another similar work proposed a position mooring system designed for a barge using ropes/wires, wherein the motion controller and control allocation systems were unified [50]. The position and orientation of the vessel were controlled by changing the tensions of the mooring lines with the sub-winch. This demonstrates that the proposed design method is more efficient than traditional modeling and simulation methods.

In 2018, Sha and colleagues [51] investigated the impact of rich wave and wind spectra on the structural integrity of a bridge spanning a deep fjord, employing a time domain approach. While addressing nonlinear responses that are challenging to resolve in the frequency domain, they calculated frequency domain transfer functions using WADAM, a software designed for frequency domain analysis of stationary vessels. Notably, their study prioritized structural response over real-time simulation performance.

Our team at NTNU has been actively engaged in virtual prototyping (VP) research since 2014, collaborating with maritime industrial partners on several intriguing projects. In one study, Chu and Zhang [42] utilized a VP framework to address mechanical, hydraulic, and control system design challenges in a crane application. They also explored advanced visualization techniques to simulate vessel movement under regular sea conditions using sinusoidal waves. This research delves into the complexities of virtual prototyping for multi-domain systems, highlighting challenges related to modern engineering system design, multi-domain dynamic system modeling and simulation, as well as data exchange and visualization. Concurrently, our team investigated ship model, simulation, and control design using VP through an XML- and TCP-based simulation bus. This approach, although effective, involves a laborious and error-prone configuration process and exhibits slower execution times [52]. Through a case study focusing on trajectory tracking, the framework demonstrated efficiency in constructing scenarios and testing control models for ship maneuvering applications.

After that, in collaboration with Subsea7 and the Offshore Simulation Center (OSC), a real-time virtual prototyping (VP) model was developed to simulate the entire process of pulling a water injection flexible riser from an installation vessel to a jacket platform, as part of a riser replacement task [53]. The focus of this research was on analyzing the configuration, tension, and maximum bending curvature along the flexible riser during the operation. In another noteworthy case study, conducted in 2019, a VP approach was employed to analyze positioning systems for drilling operations in the Barents Sea [54]. The study introduced a framework for comparing three different positioning solutions for mobile offshore drilling units using high modulus polyethylene (HMPE) ropes: active mooring with an HMPE rope, conventional dynamic positioning (DP), and active hybrid position-keeping.

Additionally, a brief summary of the literature review on VP is provided in Table 2.1.

After reviewing some literature on VP, we could get the following conclusions [55].

- **Flexibility and extensively**: Enable flexible modification and reuse of models across various simulation scenarios, facilitating rapid simulation and visualization of conceptual designs for new product and system development through modeling. [9].
- **Integration complexity and difficulty**: Address the challenge of integrating maritime product design and operation as a whole, which is not as straightforward as simple arithmetic; it involves complex coordination across multiple disciplines.
- **Necessity of a general simulation framework**: Provide an integrated platform for simulating and co-simulating dynamic models, including managing parameter and variable specification, model exchange, and selection of variables, domain models, and meta-models to ensure compatibility.
- **Still one-way communication**: Typically used during the design and concept validation phases due to the lack of closed-loop communication between models and their physical counterparts, albeit with certain limitations.

2.3 Digital Twins

In essence, the digital twin should be able to take advantage of all digital information available for an asset: system and data information models, 3D visualization models, mathematical models, dependability models, condition and performance indicators and data analysis. An increasing number of systems and processes on board a modern vessel, are dependent on computers and networks for monitoring and control. Feedback loops from system measurements are included in the computations and affect the controllers. As these systems exploit the benefits of increased connectivity and various sensors, the complexity level is rising and new methods for design,

Table 2.1 Virtual Prototyping related literature review

No.	Document	Title	Author keywords
1	[46]	Virtual prototyping of a floating wind farm anchor during underwater towing operations	Marine simulator, Virtual prototyping, Underwater towing operation, Anchor, Floating wind farm
2	[47]	A mathematical model of virtual simulation for deepwater installation of subsea production facilities	Virtual reality, Installation simulation, Mathematical model, High reality
3	[48]	A virtual reality simulation for coordination and interaction based on dynamics calculation	Virtual reality, Collaborative operation, HMI, Dynamics calculation, Offshore lifting and installation
4	[49]	Towards, Marine dual fuel engines digital twins–integrated modelling of thermodynamic processes and control system functions	Marine dual fuel engines, 0D/1D simulation, Control system functional modelling, Operating modes switching, GT-ISE modelling, Digital twins
5	[50]	A Study on position mooring system design for the vessel moored by mooring lines	Boats, Resource management, Winches, Actuators, Mathematical model
6	[51]	Numerical investigations of the dynamic response of a floating bridge under environmental loadings	Floating bridge, Environmental loads, Time-domain analysis
7	[42]	Virtual prototyping for maritime crane design and operations	Virtual prototyping, Maritime crane, Dynamics, Hydraulics, Visualization, Co-simulation
8	[52]	Towards a Virtual prototyping framework for ship maneuvering in offshore operations	Virtual prototyping, Ship maneuvering, Modeling and simulation, Plug-in control mechanism
9	[53]	Flexible riser replacement operation based on advanced virtual prototyping	Flexible riser, Replacement operation, Virtual prototyping, Finite element method
10	[54]	Virtual prototyping: a case study of positioning systems for drilling operations in the barents sea	Barents sea, Environmental cost estimation, Framework for simulation integration, Hybrid mooring position keeping, Time domain simulation, Virtual prototyping

testing, and verification are required. A major advancement in digital twin development is the integration of physics-based and data-driven approaches. This integration allows for the creation of digital twins that combine data-driven models, which are based on empirical data, with physics-based models, which are based on physical laws and relationships. The combination of these two approaches provides a more accurate representation of the system behavior, taking into account the strengths and weaknesses of both approaches.

The use of digital twins has been facilitated by the development of cloud-based simulation platforms. These platforms provide access to high-performance computing resources, allowing for the simulation and analysis of complex and large-scale dynamic operation systems. This has made the development of digital twins more accessible and cost-effective, enabling smaller organizations to take advantage of these technologies. In addition, the integration of digital twins with optimization techniques has opened up new opportunities for enhancing dynamic operation systems. Digital twins can be used to identify the optimal operating conditions for a system, taking into account multiple objectives and constraints. This has made it possible to find the most efficient and effective solutions for complex dynamic operation systems. Finally, the integration of digital twins with machine learning and artificial intelligence has provided new opportunities for the analysis and optimization of dynamic operation systems. Machine learning algorithms can be used to identify patterns and relationships in simulation data, allowing for the optimization of dynamic operation systems and the reduction of simulation time.

For production and manufacturing, the advent of Industry 4.0 poses new challenges to the next generation of smart manufacturing that should enable high adaptability and sustainability, rapid design changes, digital information technology, and more flexible training of skilled workers [56]. Digital twin technology is considered as one of the key technologies to achieve these goals. Several papers [56–60] have studied the current state of digital twin technology in manufacturing in terms of smart manufacturing [61, 62], assembly [63], modeling and operation [64], and predictive maintenance [65]. Towards smart manufacturing, a new concept of digital twin shop (DTS) based on digital twins and its key components was explored [61]. In addition, the authors studied the operation mechanisms and implementation methods of the DTS and examined the key technologies and future challenges, respectively. To enable predictive maintenance of manufacturing resources, a method for calculating the remaining useful life (RUL) of mechanical equipment using the digital twin concept was introduced in [65]. Their proposed method was validated by a case study of predicting the RUL of an industrial robot. The automotive, aviation, and maritime industries share similar common characteristics and are closely linked to manufacturing, and have recently attracted attention following the boom in the application of digital twin technology in manufacturing.

The automotive industry has changed dramatically over time as consumer demand has grown and new technologies have entered the industry [66]. In the automotive industry, digital twin technology is widely used in different stages of design [67–69], assembly [70–72], manufacturing [73–75], and maintenance [76–78]. In terms of design stage, one research work proposed a framework that uses digital twin stations to minimize the time required to develop and design a new assembly line [68]. In their study, a digital twin of a real station of an automotive assembly line was proposed to validate the proposed numerical procedure and design methodology. Xie with other colleagues developed an adaptive development environment based on digital twins for automotive cyber-physical systems (CPSs), addressing the key issues of the need to clone a corresponding digital twin for each physical entity and the need for close interaction between the digital twin and the physical entity [79].

Through their case study with an automotive body control system, it is demonstrated that the proposed adaptive development environment has low complexity and cost, high scalability, adaptability, and flexibility to meet various requirements during the development process. The share of electric vehicles has tended to expand over the past decade, which indicates that digital twin technology has great potential in electric vehicles [66]. A survey [80] systematically reviewed digital twin technologies for electric vehicles and classified electric vehicle systems into specific domains such as advanced driver assistance systems [81], vehicle health monitoring [82], battery management systems [83], electric power drive systems [84], and so on. In addition, digital twin technology has been applied to drivers in recent years. Other review and perspectives in [85] first proposed the driver digital twin (DDT) and divided the DDT architecture into four parts: real driver, digital driver, multimodal interface, and related applications. The authors summarized the key functions of DDT and established a framework for DDT, such as driver distraction detection, attention estimation, intention inference and prediction, drowsiness detection, and driver emotional state monitoring.

As a field that requires highly safe and reliable systems, the aviation industry is challenged by high production and maintenance costs, digitization, and cutting-edge technologies. Digital twin is considered as one of the technologies that can solve these issues. In the aviation industry, digital twin is applied in the whole life cycle perspectives of design [77, 86, 87], assembly [88–90], manufacturing [91–93], and operation and maintenance [94–96]. Among them, health diagnosis [97, 98] and life prediction [95, 99] are commonly studied topics. In [100], a versatile probabilistic model for diagnosis and prognosis was developed, which used the concept of dynamic Bayesian networks (DBN) to realize the vision of a digital twin. The proposed approach was illustrated by an example of fatigue crack growth in an aircraft wing. The literature [95] investigated an aero-engine predictive maintenance framework driven by digital twins and validated the model by consistent evaluation of virtual and real data assets. Through a case study of an aero-engine, their approach was proven to be effective and the model predicted with high accuracy. Xiong and Wang provided an overview of the state-of-the-art digital twin technologies in the whole life cycle of aviation system and indicated that aviation digital twin technologies are most widely used in the manufacturing and maintenance phases [101]. In addition to traditional aviation, digital twin technology for unmanned aerial vehicles (UAVs) has also received much attention in recent years, with applications in bridge inspection [102], post-earthquake construction [103], smart cities [104], and other fields.

The application and research of digital twin technology in the maritime industry lags quantitatively behind its application in the automotive and aviation industries. However, in recent years, research on digital twin technology in the maritime industry has matured as the number of related published works has increased exponentially. Digital twin technology is used in different phases of the maritime industry, including design [105–107], manufacturing [108–110], operation [111–113], maintenance and service [114–116], or multiple phases [117–119]. [120] studied and concluded systems in the maritime industry that often use digital twin technology, including ship electrical systems [121], propulsion systems [122], hull/structure [123], ship diesel engines [124], and ship behavior [125]. [114] built a data-driven digital twin ship that utilizes a large amount of information collected from sensors on board and is used to estimate the speed loss due to marine fouling. The research in [113] introduced an innovative concept of digital twins to predict ocean waves and hydrodynamic performance, which allows risk and optimal routing to be predicted in real time. To implement the proposed real-time digital twin, they developed a sophisticated wave reconstruction algorithm for real-time wave prediction and an analysis program to solve the coupled seagoing-maneuvering problem. They believe that by extending this approach, it is possible to predict the risk and performance of ships in a variety of marine environments. Mauro and Kana [126] conducted a critical systematic review of digital twin technologies used for ship life cycle and based on their paper categorization and analysis, they state that research should focus on topics related to the design and decommissioning phases, especially regarding the design of future ships. In addition to ships, digital twin technology has been applied to offshore oil and gas facilities [127], offshore wind turbines [128], and underwater drones [129].

Even though significant progress has been made in literature, for example as shown in Table 2.2, the technology of DT in maritime applications is still a challenging topic which should receive special attention by research society and industry.

- **Flexibility and multi-purpose requirements**: As previously discussed, digital twins hold significant potential for enhancing the life-cycle services within the maritime industry. Therefore, it is imperative that the models and sub-models exhibit flexibility and adaptability across various life cycle phases. Digital twins should offer an appropriate level of fidelity for each phase, ranging from basic fidelity for design to high fidelity for control and verification, and real-time capabilities for operational phases. However, current research on digital twins often addresses only specific life cycle phase requirements rather than the entire value chain. Moreover, in the maritime industry, striking a balance between the number of defined domain models and the scale of the entire system is crucial to ensure the flexibility and efficiency of the digital twin system. Consequently, there remains a need to enhance model flexibility and adaptability.
- **Limited complexity and extensibility**: Similarly, in the context of virtual prototyping (VP), current digital twin research in maritime applications tends to focus on relatively small modules, systems, or individual components. However, vessels comprise a multitude of systems and sub-systems, spanning various disciplines such as mechanical, thermal, hydraulic, and control. It is essential to integrate all significant sub-systems, including sensors, data mining, main propulsion systems, hydrodynamics systems, deck machinery, and control systems. Despite this, existing literature presents limited research on such comprehensive digital twin systems.
- **Great potential to support maritime operation**: Digital twins are not only valuable for design purposes but also play a crucial role in demanding operations, particularly for offshore vessels operating in harsh environments characterized by winds, waves, currents, and low temperatures. These operations necessitate meticulous planning and pre-execution training. In this regard, a digital twin of the entire ship can serve as a valuable tool for pre-operation planning. It enables iterative

Table 2.2 Digital Twin related literature review

No.	Document	Title	Author keywords
1	[61]	Digital twin shop-floor: a new shop-floor paradigm towards smart manufacturing	Production, Manufacturing, Convergence, Data models, Aerospace electronics, Optimization, Cloud computing
2	[64]	The modelling and operations for the digital twin in the context of manufacturing	Digital twin, Enterprise information integration, Manufacturing operations and control, Inter-enterprise interoperation, Aerospace structural parts
3	[70]	Digital human and robot simulation in automotive assembly using siemens process simulate: a feasibility study	Digital twin, Human, Robot, Automotive assembly, Tecnomatix
4	[67]	Design and simulation of a capacity management model using a digital twin approach based on the viable system model: case study of an automotive plant	Capacity planning, Digital twin, Forecasting, Viable system model, Simulation, System dynamics, Customer demand, Demand planning, Automotive
5	[81]	A digital twin paradigm: vehicle-to-cloud based advanced driver assistance systems	Digital twin, Merging, Servers, Connected vehicles, Real-time systems, Advanced driver assistance systems
6	[91]	Building a digital twin for additive manufacturing through the exploitation of blockchain: a case analysis of the aircraft industry	Blockchain, Digital technology, Additive manufacturing, Digital twin, Aircraft industry
7	[100]	Digital twin-driven aero-engine intelligent predictive maintenance	Individual aircraft tracking, Digital twin, Dynamic Bayesian network, Particlefilter, Diagnosis, Prognosis, Fatigue crack growth
8	[95]	Towards a virtual prototyping framework for ship maneuvering in offshore operations	Digital twin, Aero-engine, Predictive maintenance, Deep learning, Data-driven
9	[114]	Data-driven ship digital twin for estimating the speed loss caused by the marine fouling	Hull and propeller maintenance, Fouling, Condition-based maintenance, ISO 19030, Digital twin, Data-Driven Models, Deep learning
10	[127]	Fatigue crack growth prediction method for offshore platform based on digital twin	Digital twin, Offshore platform, Fatigue crack growth, Gaussian process, Dynamic Bayesian network

design of operational procedures, identification of inconsistencies and errors, and exploration of "what-if" scenarios during onshore planning sessions based on historical Metocean records. Moreover, digital twins can be utilized in "toolbox-talks" immediately before operation execution to rehearse procedures based on current ship loading conditions and Metocean conditions forecasted in real-time. However, research on such applications remains scarce in the existing literature.

- **Dynamic and strong interaction between DT and real counterparts**: The primary distinction between digital twins and generic models lies in their specificity to and reflection of their physical counterparts, with dynamic data flow interaction. While traditional modeling and simulation are typically conducted within a single software environment for design, control, and optimization, the complexity of modern systems, such as surface vessels with multi-domain components, renders modeling everything within one software nearly impossible. Digital twins evolve alongside their real-life counterparts throughout their life cycles, incorporating sensor measurements, simulation model updates, and software upgrades.

2.4 Digitalization Agenda

After investigation of related technology, it is necessary to discuss issues related to digitization. The digital agenda is one of the pillars of the European Strategy for Growth, which proposes to better exploit the potential of Information and Communication Technologies (ICTs) to foster innovation, economic growth, and progress. It lists "Ship Intelligence" as one of the main areas through which to achieve growth. In marine system and transport, digitalization can significantly improve design, operation and management through more accurate information on operational and infrastructure conditions and the location of vehicles and/or system behavior models. Better access to and sharing of digital data (traffic, travel, vehicle, cargo, etc.) for both public and private stakeholders along the supply chain can foster seamless information flows, and open up a wide range of new business opportunities.

Digital twins integrate artificial intelligence, machine learning and data analytics with data to create living digital models that update and change as their physical counterparts' change. A digital twin is a model of a physical asset, implemented for example as a mathematical model, an information model or a visual model. Machine learning offers a way to create and update models of the physical system based on sampled data. Plenty of methods, such as regression, support vector machine, fuzzy logic, and neural network could be used, depending on the complexity of the input/output relationship and dimension of the data. In various industrial sectors, digital twins are being used to optimize the operation and maintenance of physical assets, systems and manufacturing processes. It is a formative technology for the Industrial Internet of Things (IIoT), where physical objects can live and interact with other machines and people virtually. A digital twin can be used for monitoring, diagnostics and prognostics to optimize asset performance and utilization. Therefore, complex prognostics and intelligent maintenance system platforms can leverage the use of digital twins to find the root cause of issues and improve system productivity. The basis for state estimators is a mathematical model representing the system to be observed, and a set of sensor signals that updates the model in real-time. The differences between a digital twin from generic models is that they are specific to, and reflect their physical counterparts. The digital twin follows its corresponding real life twin through its life cycle, through collecting sensor measurements, simulation model updates, and software upgrades.

Regarding the maintenance concept, monitoring, diagnostics and prognostics are of high importance to optimize asset performance and utilization, by adding intelligence to the machines and being driven by fast and flexible predictions of the future. Such monitoring systems should also be aware of the changing operating regimes to dynamically select the appropriate prognostics model to ensure prediction accuracy. By the emerging engineering discipline of Prognostics and Health Management (PHM), "self-maintenance" could be achievable. PHM uses real-time and historical state information of subsystems and components to evaluate reliability, predicts the future, and enables intelligent decision-making. Current diagnostics and prognostics methods are not mature enough to neglect uncertainties due to complex systems and inadequate models. For that reason, from a research perspective, a PHM strategy combined with digital twins would be favorable due to its realistic real-time model of the current system condition, which has the possibility to enhance PHM by reducing anomalies and uncertainties. In this way, the maritime industry will benefit from digital twin technology [130] in different dimensions. These proxies of the physical world will help maritime companies in developing enhancements to existing products, operations, and services, and can even help drive the innovation of new businesses.

More importantly in current maritime industry, a balance between the number of defined domain models and the scale of the whole system will be met to make the digital twin system flexible and fast enough. Maritime systems communicate through a number of formats and protocols. The variety in protocols and in installed instruments and systems makes it difficult integrate with new systems, especially with those which require information from many different subsystems. In addition, each of these protocols must be set up with explicit knowledge of where and how to access each piece of information. Open source is a key factor to establish a cloud enabled co-simulation platform and model eco-system for collaborative and managed model sharing and time-domain system simulation. The core components should be open source and governance by the research consortium. The sub-models, apps, tools, services will be offered in the digital twins at various IP terms. In this way, the industrial partners could get first-hand help and speed up the design and operation.

2.5 Summary

This chapter gives an overview of modelling simulation, virtual prototyping, and digital twin technologies. Compared to the first two concepts, DT features dynamic and strong interaction between real counterparts and models, which makes DT more flexible and extendable. As mentioned above, digital twins are promising, but challenging in maritime domain. In the next chapters, we will introduce all necessary supporting technology to build our digital twin of the research vessel (RV) Gunnerus.

Houxiang Zhang received the Ph.D. degree in mechanical and electronic engineering from Beihang University in 2003 and the Habilitation in informatics from the University of Hamburg, Germany in 2011. He is currently a full Professor with the Department of Ocean Operations and Civil Engineering, Faculty of Engineering, Norwegian University of Science and Technology (NTNU), Norway. From 2011 to 2016, he also hold a Norwegian National GIFT Professorship on product and system design funded by the Norwegian Maritime Centre of Expertise. He has engaged into two main research areas, including control, optimization and AI application especially on autonomous vehicle; and marine automation, digitalization and ship intelligence. He has applied for and coordinated more than 30 projects supported by Norwegian Research Council (NFR), German Research Council (DFG), EU, and industry. In these areas, he has published over 300 journals and conference papers as the author or coauthor. He was nominated and selected as a member of the Norwegian Academy of Technological Sciences in 2019, and a member of Academy of the Royal Norwegian Society of Sciences and Letters in 2024.

Runze Mao received his M.S. degree from the Department of Ocean Operations and Civil Engineering at the Norwegian University of Science and Technology, Norway, in 2019, and his Ph.D. in 2022 from the Department of Mechanical Engineering and Science at Kyoto University, Japan. His research interests include human factors and digitization in marine operations, as well as the thermal management of electrochemical devices such as solid oxide fuel cells and lithium-ion batteries.

Pierre Major received his M.Sc. degree in electronics and information technology from the Swiss Federal Institute of Technology of Zürich, Switzerland (ETH) in 2005, and his Ph.D. degree in Engineering from Norwegian University of Science and Technology, in 2021. He is now the Head of Research at AugmentCity in Ålesund, Norway. His domains of interest are virtual prototyping of demanding offshore operations and graphical digital twins of systems such as cities or ships.

Guoyuan Li received the Ph.D. degree in computer science from the Department of Informatics, Institute of Technical Aspects of Multimodal Systems, University of Hamburg, Hamburg, Germany, in 2013. In 2014, he joined the Department of Ocean Operations and Civil Engineering, Intelligent Systems Laboratory, Norwegian University of Science and Technology (NTNU), Ålesund, Norway, where he is currently a Professor in ship intelligence. He has published more than 150 articles in the areas of his research interests, including the modeling and simulation of ship motion, autonomous navigation, intelligent control, optimization algorithms, and the locomotion control of bioinspired robots.

Hans Petter Hildre is professor and head of the Department of Ocean Operations and Civil Engineering at the Norwegian University of Science and Technology (NTNU). His area of interest is product design and system architecture design. Hans Petter is the director for Centre for Research Driven Innovation (SFI-MOVE) within marine operations. This is cooperation between NTNU, SINTEF, University Sao Paulo and 15 companies at the west coast of Norway. Professor Hildre is head of research in national program Global Centre of Expertise Blue Maritime, project leaders in several research projects, member of the board in 5 companies, and has a number of patents.

Chapter 3
Co-simulation Mechanism as Digital Twins Platform

Lars Ivar Hatledal and Houxiang Zhang

Complex heterogeneous systems, such as vessels, may require components from several different domains, perhaps developed in separate domain-specific tools. Co-simulation refers to a simulation practice, where different self-contained sub-systems are coupled to realize a more complex system. Each sub-system is a simulator that can be characterized as a black-box capable of exhibiting behavior, consuming inputs, and producing outputs [131].

Co-simulation is often used in engineering and science applications where it is necessary to model complex systems with multiple interacting components. Further, it can be particularly useful in situations where one model is not capable of representing the entire system, or where a monolithic model would be deemed too computationally expensive to run. Key properties affiliated with co-simulation are illustrated in Fig. 3.1.

Modeling a cyber-physical system (CPS) rooted in the maritime domain likely consists of components from several different tools as well as domains [18]. The plethora of modeling tools within different disciplines exists for very good reasons, and there might never be a single tool that is suitable for every stage and every branch of the design process [42]. Figure 3.2 illustrates a possible co-simulation scenario for a vessel, which requires models from several different domains. Co-simulation is essential for this scenario to succeed, not only because models from different domains need to be coupled, but also because the models may originate from different, perhaps competing companies. Due to potential concerns around exposing industrial property rights (IPR) [132], these companies might be unwilling to share models in any other form than a black-box. As the business logic can be distributed as binary code, the black-box approach should relieve this concern somewhat. However, additional resources bundled within the model may not be protected, so this approach may not completely hide IPR. On the other hand, it is easy to interface with black-box models that use standardized interfaces. Thus, it is quite manageable to establish a remote

H. Zhang et al., *Digital Twins for Vessel Life Cycle Service*,
https://doi.org/10.1007/978-981-97-4768-9_3

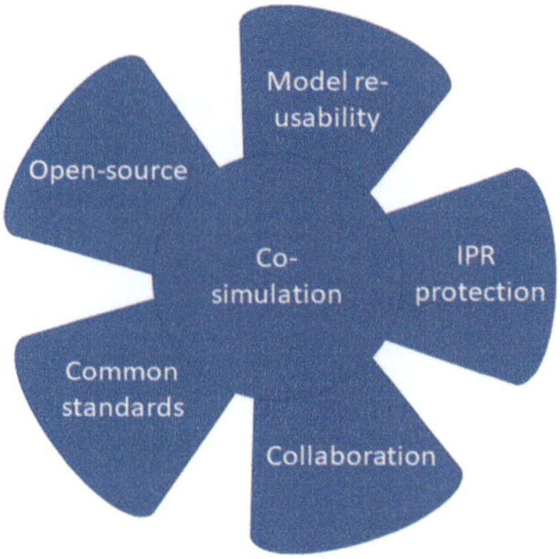

Fig. 3.1 Key properties of co-simulation

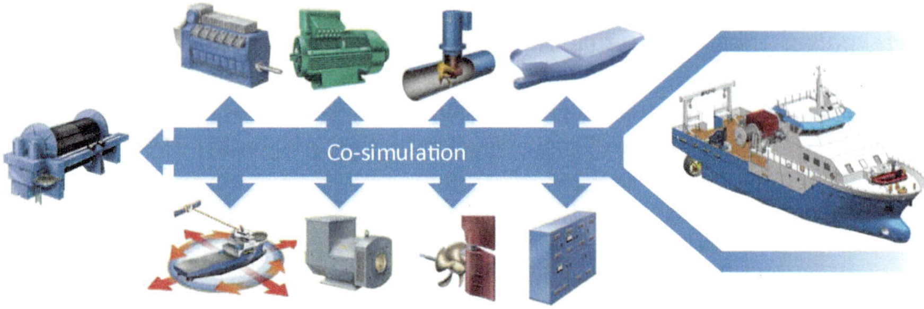

Fig. 3.2 The vessel model depicted in the figure is an aggregate of several different sub-components, which are integrated and solved together using co-simulation. *Vessel sub-component figures courtesy of the Virtual Prototyping of Maritime Systems and Operations project (Research Council of Norway, grant nr. 225322)*

interface to such models, which in turn makes it possible to hide unintentional IPR leaks by only exposing the application programming interface (API), and not the resources within.

Co-simulation can be expanded into the realm of digital twins by feeding sensor data measured from the real world into the models, which in turn closes the loop by providing actionable feedback. A crucial attribute of a digital twin model is that, since it mimics its physical counterpart, it can approximate a vessel's performance through simulation before running tests in the real world [39]. This not only provides flexibility, but also has the potential to reduce costs to a significant degree. Performance, shipbuilding, and maintenance are three vital areas where a digital twin acts as a valuable asset for a vessel [133].

3.1 Challenges of Co-simulation

Co-simulation's primary advantage lies in its ability to integrate simulation models derived from various sources, i.e. different simulation tools. Compared to more traditional monolithic simulations, where simulation models are created and executed within the same tool, co-simulation encourages re-usability, model sharing, and fusion of simulation domains. Thus, the idea of using co-simulation as a mechanism to simulate inherently multi-domain CPSs like maritime vessels and equipment is compelling.

Co-simulation is, however, not a perfect solution to overcome all simulation challenges. In [131], the authors do a good job identifying common challenges related to co-simulation; like semantic adaptation, modular coupling, stability and accuracy, and finding a standard for hybrid co-simulation. A common issue with co-simulation is related to stability and accuracy [39, 134], in particular when strongly coupled models are involved. More advanced co-simulation algorithms with error-control, which are able to roll back and retry the simulation step from a previously known stable state using a lower time-step, may mitigate this issue [135]. However, models often do not support such capabilities. A pragmatic solution to handle the accuracy issue is to lower the time-step used to drive the simulation forward, at the expense of run-time performance. Unfortunately, this is not always sufficient and the models need to be adapted by introducing elements such as spring-dampers systems at connection points to soften a stiff system. However, adapting the models naturally alters the characteristics of the intended simulation and thus it may not be an optimal solution. System designers then must re-think which parts of the overall simulation may be split into co-simulation units, making sure to avoid splitting parts of the simulation that are deemed too tightly coupled [42].

Other challenges related to co-simulation include:

- To achieve trustworthy, reliable, and relevant simulations, one naturally needs access to sufficiently good and relevant models, which might not be readily available, or they may need to be adapted for co-simulation.
- Acquired models need to be coupled. This process is often a source of error even if the models themselves are working as intended [136], especially if documentation is inadequate or missing altogether [137]. Model interfaces often expose a large number of variables, which, when coupled with non-intuitive variable names, makes the process of connecting variables difficult to reason about.
- When designing co-simulations, there might be issues related to causality when two models, which theoretically would be a good match for coupling, have inputs and outputs flowing in the wrong direction compared to each other [39, 131]. This can be alleviated by proper communication between model designers, but there might still be an issue when using legacy models.
- Naively connecting inputs to outputs between black-boxes does not necessarily imply that the resulting behavior mimics the actual couplings of the sub-system models [137]. Co-simulation often leads to less accurate simulations than monolithic ones, as each sub-system is solved individually. This changes the eigenvalues

of the total system, which can cause instabilities in the local solvers [134]. This can be alleviated to some extent by good orchestration algorithms, but such algorithms might not always be available, nor applicable due to lacking model implementations or real-time requirements.
- Challenges associated with the development of digital twins relate to insufficient synchronization between the physical and the digital world to establish closed loops, a lack of high-fidelity models for simulation and virtual testing at multiple scales, lacking uncertainty quantification for such models, difficulties related to the prediction of complex systems, and challenges related to the gathering and processing of large data sets [138].

However, co-simulation also offers a wide array of benefits that make it an indispensable technique for engineers and researchers dealing with complex systems. Some of the key advantages include:

- Co-simulation allows models generated from different tools and designed for different domains to be coupled and simulated together [33]. Each black-box incorporates its own simulation algorithm, which is usually the most appropriate for its domain [137]. This enables heterogeneous simulations to be realized, including complex CPSs like vessels [139].
- Co-simulation allows competing companies to share models towards realizing a common goal, without having to expose IPR thanks to the black-box modeling approach [39, 132, 137].
- The use of open standards encourages re-usability and collaboration [137], which saves cost and drives innovation [136].
- Co-simulation enables simulation-based commissioning of vessels and virtual sea trials to remove design flaws and implementation errors at an early stage [39].
- The inherent property of distribution-friendliness makes it viable to execute the model on a remote resource, potentially protecting the user from malicious attacks and further protecting the model owner's IPR by not physically re-distributing the model [132]. This also allows for large-scale parallel simulations [140, 141].
- Co-simulation naturally enables the realization of digital twins due to the inherent distributed and component-based structure as it allows models using different runtime systems to be executed together [142].

3.2 The Functional Mock-Up Interface

Co-simulation is a generic term that refers to the computational technique for executing simulation models, however, in recent years the term has become synonymous with the Functional Mock-up Interface (FMI) [33], a highly adopted tool-independent industry standard for *co-simulation* (CS) and *model exchange* (ME). Since its initial release in 2010, the FMI has become the de-facto standard for co-simulation in the engineering community. Today, it is widely used in industries such as automotive,

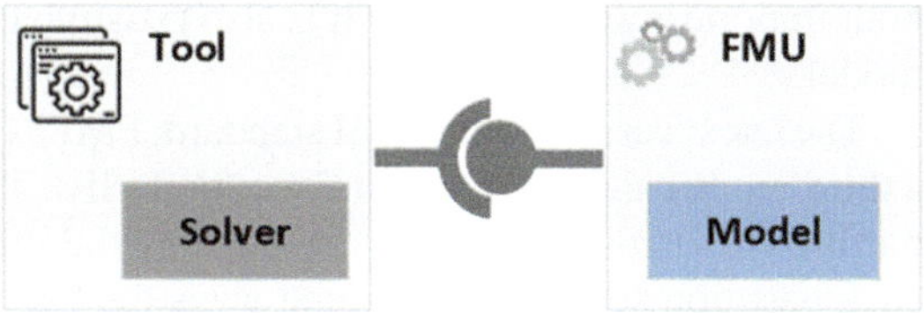

Fig. 3.3 FMI for model exchange

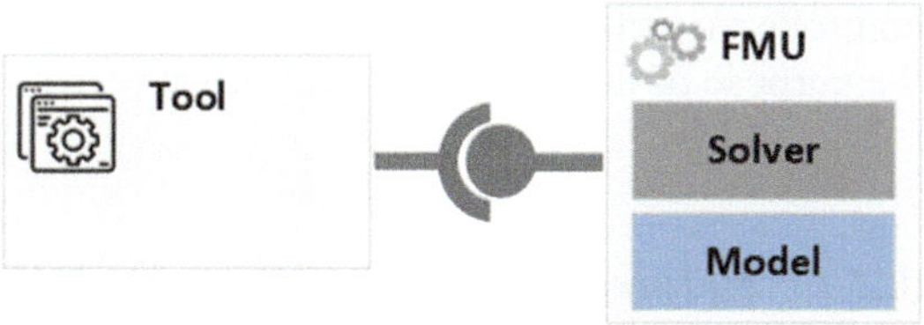

Fig. 3.4 FMI for co-simulation

aviation, energy, manufacturing, and also maritime, where complex systems are often analyzed and simulated using a wide variety of tools from different vendors. The FMI enables interoperability between these tools, allowing engineers and researchers to combine models from various domains and perform integrated simulations. The main goal of the FMI standard is to enable sharing of simulation models across tools. To accomplish this, the FMI relies on a combination of XML-files and compiled C-code packaged in a compressed ZIP archive. Up until version 2.x, the FMI consisted of two main parts, both of which a single FMU may support:

- *FMI for ME:* Models are exported without a solver, as illustrated by Fig. 3.3, and are described by differential, algebraic, and discrete equations with time-, state-, and step-events.
- *FMI for CS:* Models are exported with a solver, as illustrated by Fig. 3.4, and data is exchanged between sub-systems at discrete communication points. In the time between two communication points, the sub-systems are solved independently from each other.

A model implementing the FMI standard is known as an FMU, and may support both or either of the CS and ME modes. The key difference between the two relates to how importing tools makes the FMU step forward in time. With a CS FMU, the numerical solver is embedded. The importing tool will set the inputs, tell the FMU to step forward in time, and then read the outputs. In the case of a ME FMU, the numerical solver must be provided by the importing tool. The FMU provides functions to set the state and inputs, and to compute the state derivatives. The solver in the importing tool will determine what time steps to use, and how to compute the state at the next time step. CS FMUs contain all code necessary to abstract away the details of their internal computations. This simplifies the importer compared to Model Exchange and Scheduled Execution, at the cost of reduced flexibility of use. In practice, the CS kind is more commonly used as it is more widely supported in both importers and exporters. It is also possible to re-package a ME model into a CS model.

The latest version of the FMI standard, FMI 3.0 [143], released in 2022, introduces a third mode called Scheduled Execution (SE). This mode exposes individual model partitions, where a scheduler provided by the importer can control the execution of each partition separately. In some ways the SE interface has similarities to the ME interface. The SE interface externalizes a scheduling algorithm usually found in a controller algorithm, while the ME interface externalizes an ODE/DAE solver. For the purpose of this chapter, however, we focus on FMI for CS.

3.2.1 Related Simulation Standards/Interfaces

In the following, a number of standards for co-simulation is presented that are highly related or connected with the FMI.

Distributed Co-simulation Protocol

The distributed co-simulation protocol (DCP) is a protocol used to facilitate co-simulation between different simulation tools or models in a distributed environment [144]. It provides a standard interface for exchanging simulation data and control messages between simulation components, allowing them to interact with each other in a coordinated and synchronized way. The DCP is compatible with the FMI standard and relies on a standard network interface rather than C-code for portability. The DCP, like the FMI and the SSP, is maintained by the Modelica Association.

Digital Twin Assembly

Digital twin assembly (DTASM) is a binary interface (ABI) for executable numerical simulators compiled into WebAssembly modules [145]. WebAssembly (WASM) is an emerging technology, first announced in 2015, that offers an alternate portability route than the traditional C interface approach used by e.g., the FMI. It is a low-level binary format and virtual machine designed to be a portable target for the compilation of high-level languages. WASM is intended to be a safe, fast, and portable format that can be executed on any platform that has a WASM interpreter, which includes most modern web browsers as well as standalone runtimes outside of the web. DTASM leverages WASM, providing a language-independent binary interface that is portable and sandboxed, which is functionally comparable to FMI for co-simulation but not tied to a particular programming language. Overall, this is an interesting approach that should be closely monitored.

Asset Administration Shell

The asset administration shell (AAS) is a standardized information model that is being developed as part of the Industry 4.0 initiative [146]. It is a digital representation of a physical or virtual asset, such as a machine, product, or component. The concept is closely related to the concept of digital twins. In fact, the AAS can be seen as a key component of a digital twin. The AAS provides a standardized way to describe the properties, capabilities, and behavior of an asset, which can be used to create a digital twin that accurately represents the real-world asset. This enables the digital twin to communicate and interact with other assets, systems, and processes in the same way that the physical asset does. The FMI is one of the technologies used within the AAS framework.

By combining the AAS with digital twin technology, it is possible to create a comprehensive, end-to-end digital representation of an asset that spans its entire life cycle, from design and manufacturing to operation and maintenance. This can enable a range of benefits, such as improved asset performance, reduced downtime, and more efficient maintenance and repair. Overall, the combination of the AAS and digital twin technology is a powerful enabler of the digital transformation of industry, and is expected to play a key role in driving innovation and competitiveness in a wide range of sectors.

3.2.2 Related Structural Standards

Co-simulation often deals with the simulation of multiple, likely interconnected, models. The FMI standard itself only deals with the definition of a single model. However, other supplementary standards exist for handling the definition and packaging of an assembly of models following the FMI standard. These standards are important companions for the FMI as they make simulations easier to reason about and transfer between entities and tools.

System Structure and Parameterization

As an extension of the FMI standard, the SSP [147] standard aims to facilitate the design, simulation, and execution of a network of components (e.g., FMUs). It is a tool-independent standard to define complete systems consisting of one or more components, including its parameterization, which can be transferred between simulation tools. The structure of a simulation is configured through XML, which defines which models make up a simulation (components), which variables are exposed (connectors), how they are connected (connections), and how they are parameterized (parameter-sets). Additionally, annotations may be present that are used for defining tool-specific features. The configuration is bundled, together with any resources required, like FMUs, into a compressed ZIP archive. The main use of this standard is to define:

1. A standardized way to store and apply parameters to a set of components.
2. A standardized format for representing the interconnections between a set of components.

Thus, the SSP allows complete systems to be defined, shared, and re-used in a tool-independent manner. Therefore, the SSP helps empower the users so that they in large parts may choose for themselves, and make informed decisions, through testing, in which importing tool they opt to use for their simulation needs. Much as the FMI allows decoupling from the simulation tools, the SSP allows decoupling from the co-simulation *orchestrator* (see Sect. 3.3).

Open Simulation Platform—Interface Specification

The OSP Interface Specification (OSP-IS) [148] is an addition to the FMI 2.0 standard for co-simulation, which provides a method for adding semantic meaning to model interface variables. The OSP-IS aims to facilitate faster and less error-prone construction of co-simulation systems by simplifying the model connection process, and validation of semantically correct simulations. Basically, the OSP-IS allows model creators to bundle an additional XML document alongside an existing FMU, which describes the available connections using a standardized ontology.

Open Simulation Platform—System Structure

The OSP system structure (OSP-SS) is an alternative to the SSP standard, developed as part of the OSP. Like the SSP, it is based on XML and provides similar features. A major difference lies in how connections are defined. The OSP-SS natively supports OSP-IS connections, allowing groups of variables to be connected between models that support it. Additionally, it enables the declaration of functions. Functions in this context are standardized code blocks that are placed between FMUs to transform their signals. This allows more advanced variable transformations than the few generic mapping functions found in the SSP, but requires more effort to support. Along with the OSP-SS, the OSP project also provides a graphical user interface for creating definitions conforming with the standard through a free, but closed-source, tool named *Kopl*. Like with the SSP, it is completely feasible to implement the required XML content manually. To check the validity of a configuration, a command-line tool called the *OSP-validator* is freely available.

3.2.3 *Open Libraries and Tools for FMI Based Co-simulation*

This section gives an overview of existing open-source libraries and tools supporting FMI-based co-simulation that have seen development in recent years.

Table 3.1 shows a variety of open-source software libraries that support importing FMUs. This means that the library will be able to unzip the FMU archive, parse the *modelDescription* XML and interact with the underlying C-interface—distributed as platform-specific binaries. As seen, support for the latest standard is still poor and will hopefully improve soon. Many of the libraries still support the initial version of the standard for compatibility reasons. Preferably, no new models should be built using version 1.0 of the standard so that importing tools at some point can opt out of supporting it.

Table. 3.2 shows a number of open-source software libraries that support exporting FMUs. Some are more high level than others and the effort required to export a model varies. For example, FMU4cpp, FMI4j, and PythonFMU all make use of cppfmu, but provide additional higher-level abstractions. Note that unifmu [149] is technically language neutral as it is based on a communication channel, but provides a ready-to-use framework for the languages mentioned. A common denominator among the tools listed is that models are written using a generic programming language. This offers great flexibility, but is not always the preferred approach. An alternative would be to use the export facility of FMI compatible simulation & modelling tools such as SimulationX, 20Sim, OpenModelica, or similar.

The tools and libraries that have been listed all deal with individual models. However, when designing systems one typically has to couple multiple models. The SSP, introduced earlier, can be used for this purpose in a tool-independent manner. To simplify creating SSP archives, our research group developed sspgen [150], a Kotlin-based domain specific language (DSL) designed for creating SSP 1.0 compatible systems. Using sspgen, SSP systems may be distributed as single, easily modifiable, and executable source file rather than zipped archives (.ssp). Instead

Table 3.1 A selection of open-source software libraries providing FMI import

		FMI support		
Name	Language	Co-simulation	Model exchange	License
FMI Library	C	v1, v2	v1, v2	BSD
FMI++	C++, Java, Python	v1,v2	v1, v2	BSD
FMI4cpp	C++	v2	v2	MIT
PyFMI	Python	v1, v2	v1, v2	LGPL
FMPy	Python	v1, v2, v3	v1, v2, v3	BSD
fmi4c	C	v1, v2, v3	v1, v2, v3	MIT
FMI4j	Java	v1, v2	v1, v2	MIT
JavaFMI	Java	v1, v2	–	LGPL
FMI.jl	Julia	v2, v3	v2, v3	MIT

Table 3.2 A selection of open-source software libraries providing FMI export

		FMI support		
Name	Language	Co-simulation	Model exchange	License
cppfmu	C++	v1, v2	v1,v2	MPL
FMU4cpp	C++	v2	–	MIT
FMU4j	Java	v2	–	MIT
PythonFMU	Python	v2	–	MIT
Reference FMUs	C	v1, v2, v3	v1, v2, v3	BSD
unifmu	Python, C#, Java	v2	–	–

of writing large, complicated XML documents defining the simulation structure, it allows the same document to be written using a simplified Kotlin DSL. This has several advantages over manually writing XML, such as the ability to compute parameter values. For example, it is possible to set the value of a parameter as the result of some function invocation, like $realParameter = PI/2.0$. Furthermore, entries may be copied and subsequently modified, leading to less verbose and error prone documents. sspgen also automatically validates the content and handles the packaging of the required XML, together with any additional resources required, into a ready-to-use SSP archive. Resources can be both local files and URLs, thus allowing FMUs to be fetched from a remote repository. Furthermore, sspgen also allows OSP-IS connections to be formed, enabling validation of semantically correct simulations as well a more streamlined model connection process. This is achieved by transforming the OSP-IS variable groups into corresponding scalar connections supported by the SSP.

3.2.4 Open Platforms for FMI-Based Co-simulation

The previous sections introduced various standards, supporting libraries, and tools for FMI-based co-simulation. However, complex interconnected co-simulations, such as simulations for demanding maritime operations, require a capable co-simulation platform to solve the overall system. The platform should be able to import all models required, connect them, and simulate them together using a suitable orchestration algorithm.

Table. 3.3 shows a selection of open-source platforms for handling co-simulations. Furthermore, co-simulations may also be carried out in commercial simulation applications like 20Sim, Dymola, SimulationX, and others. For further reference, consult the FMI standard's tools page.[1] To count as a platform in this respect, the tool must be able to solve systems of interconnected FMUs using some form of orchestration

[1] https://fmi-standard.org/tools/.

Table 3.3 A selection of open-source co-simulation platforms supporting the FMI

	FMI support									
	Co-simulation		Model Exchange							
Name	v1.0	v2.0	v1.0	v2.0	SSP	Distributed	API	CLI	GUI	License
DACCOSIM		✓				✓			✓	LGPLv3
Maestro		✓				✓	✓		✓	GPLv3
MasterSim	✓	✓					✓	✓	✓	BSD3
FMPy	✓	✓	✓	✓	✓		✓	✓	✓	BSD
OMSimulator		✓		✓	✓		✓	✓	✓	GPLv3
OSP	✓	✓		✓	✓	✓	✓	✓	✓	MPLv3
Ecos	✓	✓		✓	✓	✓	✓	✓		MIT

algorithm as described in Sect. 3.3. The FMI is flexible enough to allow both simple and advanced orchestration algorithms to be implemented. Which algorithm to choose depends on the availability of algorithms on the platform in question, as well as the capabilities of the models participating in the simulation. While more advanced co-simulation algorithms would be preferable for accuracy reasons, in practice, however, many models lack the optional capabilities that make such algorithms possible. E.g., the ability to save and restore state and acquire directional derivatives. Thus, more basic algorithms like variations of a fixed-step algorithm are often chosen.

3.3 Co-simulation Orchestration

The FMI enables interoperability between diverse models, but the standard itself does not specify exactly how such models should be orchestrated. Orchestration refers to the coordination and management of various components or subsystems to achieve a specific goal or behavior. In the context of FMI, these components are the Functional Mock-up Units (FMUs). Orchestration algorithms play a vital role in controlling the execution of FMUs, managing their data exchange, and ensuring that the simulation progresses smoothly. These algorithms enable engineers to define how different FMUs interact and how data flows between them, creating a holistic simulation environment. One of the fundamental aspects of simulation orchestration is the concept of a *step size*. The step size defines the time increment at which the simulation progresses. Orchestration algorithms must coordinate the advancement of time and ensure that all FMUs progress synchronously. This requires careful consideration of the numerical integration methods used by the FMUs to update their states. Another critical aspect of orchestration is data exchange between FMUs. Each FMU may expose inputs and outputs, and the orchestration algorithm must manage the flow of data between these variables. This includes handling data interpolation, extrapolation, and synchronization to maintain numerical stability and accuracy.

Several common orchestration algorithms are used in the context of co-simulation and the FMI:

- **Fixed-Step Solver:** The fixed-step solver is a straightforward orchestration algorithm where time advances in fixed increments, and all FMUs update their states simultaneously at each time step. This method is simple but may lead to inefficiencies when FMUs have varying computational requirements. Any FMU can be used for this type of solver.
- **Variable-Step Solver:** The variable-step solver is a more sophisticated algorithm that adapts the step size dynamically based on the needs of each FMU. It can improve simulation efficiency and accuracy by allowing FMUs to take larger steps when possible and smaller steps when necessary to capture rapid changes. These types of solvers typically require that the FMUs support optional features found in FMI 2.0 e.g. state rollback.
- **Event-Driven Simulation:** In event-driven simulation, FMUs update their states only when certain events occur, such as trigger conditions being met. This approach is particularly useful for systems with sporadic or discrete changes and can conserve computational resources. This solver type requires FMI 3.0.

These algorithms manage the intricacies of time advancement, data exchange, and synchronization, making it possible to create holistic and accurate simulations. However, their successful implementation requires a deep understanding of numerical methods, simulation semantics, and the specific requirements of the engineering problem at hand. See e.g. [135, 151] for notes on implementation details.

3.4 Creating and Running Co-simulations

Co-simulation projects can be complex and may require expertise in various domains such as modeling, simulation, and programming. Collaboration with experts and careful validation of results are essential for the success of any given co-simulation. To further aid the reader, the following provides some guidance on the steps that should be taken for setting up and running a co-simulation. Moreover, further guidelines on co-simulation can be found in e.g. [152, 153].

3.4.1 Problem Definition

In the initial stages of co-simulation, it's essential to clearly define the problem to tackle. This means thoroughly understanding the issue at hand, considering all its complexities. This helps set the stage for the entire co-simulation process, making it easier to identify the essential variables, parameters, and possible limitations for the subsequent modeling and analysis steps.

3.4.2 Modelling

A number of commercial and open-source modelling tools exist that support the FMI, providing users with the flexibility to create models in their preferred environment. Which one to choose depends on various factors, including your specific requirements, the complexity of the system, licensing constraints, and familiarity with the tool. If flexibility is needed, solutions exist that facilitate exporting models from source code, such as the previously mentioned *PythonFMU*, *FMU4j*, and *FMU4cpp* tools.

To simplify and ensure proper usage of any generated model, supplementary documentation should be bundled. Additionally, the OSP-IS should be adopted in order to simplify and allow verification of the connections in the configuration step. Insufficiently documented models are not only difficult to use, know-how may be lost over time leading to wrong or restricted usage. If possible, the source code or project files used to create the model should be bundled, or somehow linked to, in order to provide additional insights and more easily facilitate future maintenance.

3.4.3 Configuration

The configuration step is vital in order to create functional co-simulations. The SSP standard can be used for this purpose, allowing the system to be imported by more than one tool. Steps to configure the simulation using an SSP workflow may be as follows:

1. **Collect FMUs:** Start by collecting the FMUs to simulate. Then declare and specify initial conditions.
2. **Connect Inputs and Outputs:** Define the connections between the input and output variables of the FMUs. This involves specifying how the variables from one FMU are connected to the variables of another FMU. For large systems this may be error-prone and time-consuming task as the FMI and SSP work with individual scalar values.
3. **Configure Simulation Settings:** Some execution frameworks require or allow defining simulation properties, like solver settings, as part of the SSP configuration. These settings determine how the co-simulation progresses in time. Adjust the solver settings and similar according to the requirements of your simulation.
4. **Define Simulation Scenario:** This step specifies the conditions under which the co-simulation will run. Some execution frameworks allow setting up *scenarios* through tool-specific configuration files that define the simulation scenario, including input signals (if any), and any other parameters relevant to the co-simulation.

To simplify and verify the configuration of the SSP system to be simulated, *sspgen* could be used, which also facilitates the use of the OSP-IS in an SSP context.

3.4.4 Execution

With the models and configuration in-place, it is time to perform the actual simulation. Multiple open-source and commercial frameworks and applications exist that support both the FMI and the SSP, each with its own strengths and weaknesses. Even more options exist when dealing with only FMUs, but then the system must be configured in a tool-specific way. Through the work with the Gunnerus digital twin, the authors have used the Vico (and later Ecos) and OSP tools. These tools provide both a programming API and a command-line-interface. For fine grained control, using a programming API is advised. To run a simulation, the user would provide the SSP to the simulation system and specify some orchestration algorithm and a step-size. These settings would influence the speed and accuracy of the simulation.

Sandboxing

Blindly running FMUs poses a significant security threat and steps should be taken to alleviate this issue. Sandboxing is a security mechanism that isolates running programs and processes from the rest of the system, preventing them from affecting other programs, system resources, and data. It provides a controlled environment where applications can be executed, tested, or analyzed without posing a threat to the operating system or other software running on the same system. Here are a few scenarios where sandboxing might be relevant in the context of co-simulation and the FMI:

1. **Security Concerns:** The FMI allows integration of simulation models from various tools and sources. When dealing with simulation models from untrusted sources or integrating models from external parties, sandboxing could be employed to isolate these models to run them securely within the simulation environment without compromising the stability and security of the overall simulation system.
2. **Testing and Development:** When developing or testing FMI-based simulation models, sandboxing techniques could be useful. Isolating the simulation environment can help in identifying runtime requirements and other issues. Furthermore, when testing different scenarios, we can ensure that the execution won't interfere with the rest of the simulation system.

Sandboxing of co-simulations can be achieved by running the entire simulation in a virtual environment. Even in this case, there is a potential issue that the FMUs can access the data contained in other models participating in the simulation. To further isolate the models, each model or part of the simulation may be isolated using tools like *proxyfmu* or *unifmu* that make it possible to integrate models running on another system. DNV provides a commercial solution to this issue through their Simulation Trust Center based on the OSP tools, including *proxyfmu*.

3.4.5 Ecos Co-simulation Platform

To support this research, the authors have developed *Ecos*, a high-level co-simulation framework written in C++ that is also accompanied by interfaces for C and Python. The framework is based on the lessons learned through the development of *Vico* [16], a Java-based co-simulation framework as well as the OSP initiative. Compared to the libcosim library from the OSP, Ecos is more research-oriented. Some of the features of Ecos are:

- Bundles a C-API for easier integration with other applications.
- Support for both version 1.0 and 2.0 of the FMI standard for CS.
- Support for version 1.0 of the SSP standard, which allows complete simulation systems to be represented in a standardized way.
- Bulk read/write and caching of variable data for efficient access.
- Integration with proxy-fmu, allowing distributed execution of FMUs. This in turn enables models to be run regardless of platform, license and software dependencies (with limitations).
- The ability to specify events, inline or through configuration files, to occur at specified trigger points, through so-called *scenarios*.

Like *Vico* and the OSP tool *libcosim*, Ecos has adopted the use of *proxy-fmu*. proxy-fmu is a C++ library based on work presented in [154] that allows distributed execution of single FMUs. In some cases, users may encounter issues when trying to run an FMU that are related to one of the following:

1. The FMU does not support the current operating system.
2. The user has security concerns regarding the content of the FMU.
3. The FMU requires a license that is not available on the target computer, but is readily available on another.
4. The FMU may not instantiate more than one model instance within a single process.
5. The FMU conforms to the FMI 1.0 specification, while the importing environment only supports version 2.0.

proxy-fmu can solve these issues by wrapping the original FMU in a server program, which is then accessed using platform-independent remote procedure calls. In this way, the FMU may either run in a sandbox or on a different computer altogether.

A crucial part of any co-simulation tool is the available orchestration algorithms. Currently, Ecos only provides a fixed-step algorithm. However, the API facilitates the creation of additional algorithms, and as time passes, hopefully more algorithms will be added. The provided algorithm, however, can be configured to run models in parallel and with individually specified step-sizes. Example code for running a SSP system in C++ and Python using Ecos is given in Listing 3.1 and 3.2 respectively.

Listing 3.1 Ecos C++ demo

```
auto ss = load_ssp(sspDir);
double stepSize = 1.0 / 100;
auto algorithm = std::make_unique<fixed_step_algorithm>(stepSize);
auto sim = ss->load(std::move(algorithm));

auto csvWriter = std::make_unique<csv_writer>("results.csv");
csvWriter->config().load("LogConfig.xml");
sim->add_listener("csv_writer", std::move(csvWriter));

sim->init();
sim->step_until(10);

sim->terminate();
```

Listing 3.2 Ecos Python demo

```
step_size = 1.0 / 100;
sim = EcosSimulation(sspDir, step_size)
sim.add_csv_writer("results.csv", "LogConfig.xml")

sim.init()
sim.step_until(10)

sim.terminate()
sim.destroy()
```

As the authors rely on open standards like the FMI and SSP, the choice of execution framework is in large decided by the end-user. Some of the FMUs described in this book are generated by commercial tools that offer FMI export, while others are generated from source code using open-source FMI exporters. To support this work, NTNU has developed open exporters for Python, Java and C++. The source of the model depends on the problem domain, and co-simulation as an enabling technology allows the various models to be integrated and simulated as a complete system.

3.4.6 Quarter-Truck Case Study

In the following, co-simulation of a simplified quarter-truck system is demonstrated. The system is illustrated in Fig. 3.5, with m_w and m_c representing the mass of wheel and chassis respectively. Both masses have a single vertical degree of freedom coupled by a linear spring-damper system representing the chassis suspension and wheel tyres. The ground profile is given as external input. The co-simulation system representing the quarter truck is comprised of three models: the *chassis* including the

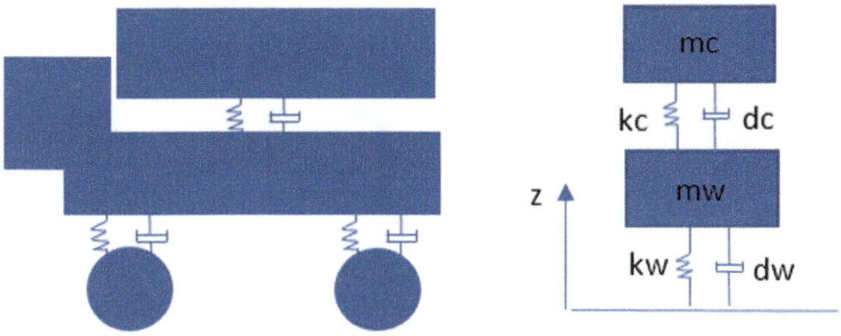

Fig. 3.5 Illustration of the quarter-truck system (from [16])

Table 3.4 Input and output variables of the quarter-truck models used for connections

FMU	Variable	Input/Output	Description
Chassis	F_{susp}	Output	Chassis suspension force applied to the wheel
	$\dot{z}_w$	Input	Velocity of the wheel from the wheel model
Wheel	F_{susp}	Input	Chassis suspension force from the chassis part
	$\dot{z}_w$	Output	Velocity of the wheel sent to the chassis part
	F_{tyre}	Output	Tyre force applied to the ground
	$\dot{z}_g$	input	Ground profile, given as vertical velocity variation from the ground model
Ground	F_{tyre}	Input	Tyre force from the truck wheel
	$\dot{z}_g$	Output	Ground profile, given as vertical velocity variation sent to the wheel

suspension, the *wheel* including the tyre and the *ground*. The system was implemented in 20Sim and the individual models exported as FMUs. OSP-IS definitions were then added to the models. The input and output variables used for connections are given in Table. 3.4. Furthermore, the sspgen definitions for the connections are the sspgen definitions for the connections are shown in Listing. 3.3. Here both scalar connections native to the SSP and FMI as well as the alternative OSP-IS type connections are shown. As demonstrated, the OSP-IS type declarations are more compact and provide more context.

$$\begin{aligned} F_{susp} &= k_c(z_w - z_c) + d_c(\dot{z}_w - \dot{z}_c) \\ F_{tyre} &= k_w(z_g - z_w) + d_w(\dot{z}_g - \dot{z}_w) \end{aligned} \tag{3.1}$$

As a benchmark for the simulation accuracy, the analytical model for the system is derived. The suspension force and the tyre force are given by Eq. 3.1, while the equations of motion for the chassis and wheel are given by Eq. 3.2.

$$\begin{aligned} m_c\ddot{z}_c &= F_{susp} - m_c g \\ m_w\ddot{z}_w &= F_{susp} - F_{tyre} - m_w g \end{aligned} \tag{3.2}$$

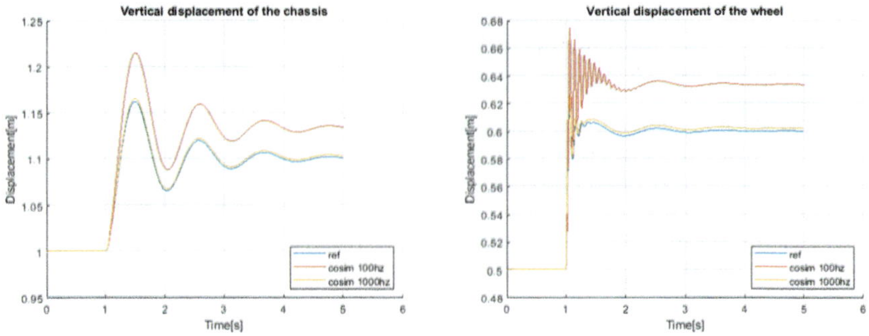

Fig. 3.6 System response for the Quarter truck co-simulation

Listing 3.3 Declaring connections using sspgen

```
// SSP type connections
connections {
    "wheel.p1.f" to "chassis.p.f"
    "chassis.p.e" to "wheel.p1.e"
    "ground.p.f" to "wheel.p.f"
    "wheel.p.e" to "ground.p.e"
}

// OSP-IS type connections
ospConnections {
    "chassis.linear mechanical port" to "wheel.chassis port"
    "wheel.ground port" to "ground.linear mechanical port"
}
```

A reference solution has been computed by means of Euler method, with the integration time step set to 0.001 s. The system response is shown in Fig. 3.6. The co-simulation results, generated by the Ecos library, are shown using both a 100hz and 1000hz fixed-step size for the orchestration algorithm. Naturally, simulating the system at 1000hz shows a clear improvement in accuracy. The increase in accuracy comes, however, with a certain run-time cost. For more details and comparison between tools, see [16].

The models and their system configuration are available from GitHub,[2] and can be simulated using any SSP compliant system. To run the system using the Ecos CLI, the command shown in Listing. 3.4 can be used. By default, Ecos will write all available variables to a CSV file. This behaviour can be controlled in more detail through an optional configuration file and users may also supply configurations to enable plotting.

[2] https://github.com/Ecos-platform/quarter_truck_demo.

Listing 3.4 Executing the Quarter-truck system through the Ecos CLI

```
ecos —path "path/to/ssp" —stepSize 0.001 —stopTime 10
```

3.5 Summary

This chapter presents co-simulation and its use as an enabling technology for our ongoing work in developing a digital twin of the R/V Gunnerus. Co-simulation stands as a powerful approach to tackle the complexities of modern engineering systems. With continuous advancements in coupling algorithms, standardized interfaces, and performance optimization techniques, co-simulation has the potential to revolutionize the way engineers analyze and design complex systems. Co-simulation allows the CPS that the vessel represents to be simulated using models from different vendors and tools. This is crucial for an aggregate model in the maritime domain, as many different vendors and domain-specific tools are usually involved. Not only does the use of co-simulation allow building of aggregate systems from different vendors, but it also allows the simulation to be performed in freely available open-source tools.

In order to elevate a co-simulation to a digital twin, real assets are needed. Like their virtual counterparts these assets can be modelled as FMUs. By having the virtual and real assets share the same underlying data interface, it is possible to swap the real and virtual models. In order to model the real asset as an FMU, however, the implementation needs to have a two-way data link with the real asset. For the Gunnerus, MQTT is used for communication. An asset that can only send data, but not receive, can still be useful and we would describe it as a digital shadow (DS).

Using the presented simulation framework, model library, and tools presented, NTNU will continue its work related to the digital twin of the Gunnerus, gradually improving its accuracy and usability. However, addressing challenges related to usability, real-time performance and consistent coupling remains pivotal to fully realizing the benefits of this technique.

Lars Ivar Hatledal received the Ph.D. degree from the Department of Ocean Operations and Civil Engineering, Norwegian University of Science and Technology (NTNU), Alesund, Norway, in 2021. He is currently working as an Associate Professor with the Department of ICT and Natural Sciences. His research interests revolves around software architecture, co-simulation and visualisation.

Houxiang Zhang received the Ph.D. degree in mechanical and electronic engineering from Beihang University in 2003 and the Habilitation in informatics from the University of Hamburg, Germany in 2011. He is currently a full Professor with the Department of Ocean Operations and Civil Engineering, Faculty of Engineering, Norwegian University of Science and Technology (NTNU), Norway. From 2011 to 2016, he also hold a Norwegian National GIFT Professorship on product and system design funded by the Norwegian Maritime Centre of Expertise. He has engaged into two main research areas, including control, optimization and AI application especially on autonomous vehicle; and marine automation, digitalization and ship intelligence. He has applied for and coordinated more than 30 projects supported by Norwegian Research Council (NFR), German Research Council (DFG), EU, and industry. In these areas, he has published over 300 journals and conference papers as the author or coauthor. He was nominated and selected as a member of the Norwegian Academy of Technological Sciences in 2019, and a member of Academy of the Royal Norwegian Society of Sciences and Letters in 2024.

Chapter 4
Data Collection and Transmission in Digital Twins

Guoyuan Li, Finn Tore Holmeset, Houxiang Zhang, and Hans Petter Hildre

Data in a variety of forms, ranging from video, audio, image, text, to sensor data, provides us with a comprehensive understanding of the research vessel. For example, the navigation equipment and resources onboard, such as Electronic Chart Display and Information System (ECDIS) and Automatic Radar Plotting Aid (ARPA), could reveal the ship status and safety level to some extent when navigating in busy waters. Data can be used not only for mining meaningful information, but also for enabling agility in an ever-changing environment and empowering data-driven decision-making. It will be demonstrated throughout the book that the use of the data plays a key role in digital twin development, from modeling and simulation (Chap. 5), to enabling tools (Chap. 6), to advanced applications (Chaps. 7, 8, and 9). To fulfil these functions, data needs to be collected continuously and transmitted to local repository or remote control center (RCC), depending on where the twin system is deployed, for training, testing, and verification. In this chapter, we will introduce what and how data is collected and transmitted, stored for the development of the twin ship system [155].

4.1 Data Infrastructure

To perceive the surrounding environment and ensure safe navigation, a variety of different sensors are deployed onboard the R/V Gunnerus. Figure 4.1 illustrates the rough location of these onboard sensors.

There are three diesel engines on the vessel. Engine data such as engine speed, temperature, and oil pressure is considered necessary information for identifying engine status in different work modes (see details in Chap. 8). Table 4.1 lists engine data we used in the digital twin system.

Taking advantage of interior sensors onboard, ship motion data could be collected, which is of great significance in reflecting ship status. The information about

H. Zhang et al., *Digital Twins for Vessel Life Cycle Service*,
https://doi.org/10.1007/978-981-97-4768-9_4

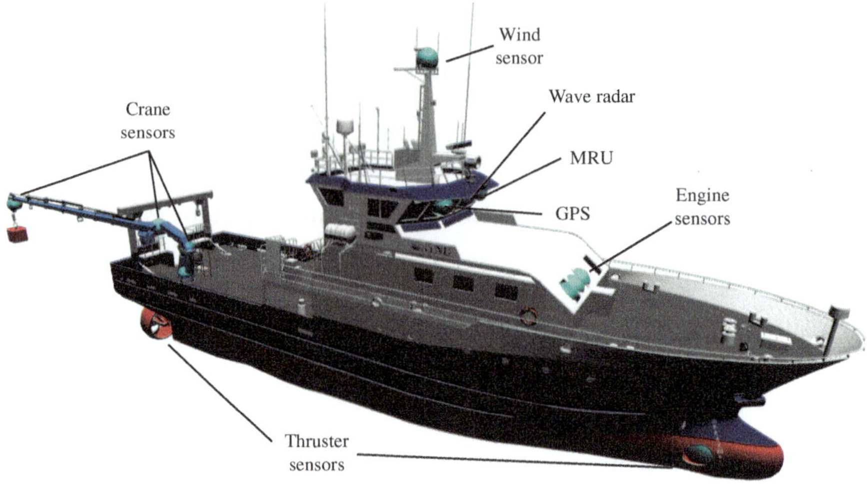

Fig. 4.1 Sensor location on R/V Gunnerus

Table 4.1 Engine data sensed onboard the R/V Gunnerus

Sensor ID	Description	Unit
1	Engine speed	RPM
2	Lube oil pressure	Bar
3	Lube oil temperature	°C
4	Cooling water temperature 1	°C
5	Cooling water temperature 2	°C
6	Boost pressure	Bar
7	Engine power	KW
8	Engine exhaust temperature	°C
9	Fuel rate	Liter/min

the two azimuth thrusters on Gunnerus (see details in Fig. 1.5), including desired and actual RPM and azimuth angle, provides detailed control information for ship maneuvering. Moreover, the data from onboard motion reference unit (MRU) contains the attitude of the vessel, such as heading, pitch, roll, and their corresponding change rate. Table 4.2 lists the ship motion data from the azimuth thrusters and the MRU. Together with the engine data, this information could help analyze navigation behavior and optimize ship voyages in the digital twin system.

In addition to these internal sensors, external sensors are also deployed on the Gunnerus vessel to help sense changes in the environment. From the global positioning system (GPS), ship position information including latitude, longitude and its course could be obtained. Moreover, the planar motion speed of the ship in the surge and sway directions, as well as ship speed over ground, is included. An anemometer and a wave radar are deployed at the top of the Gunnerus vessel, which provides

Table 4.2 Ship motion data sensed onboard the R/V Gunnerus

Sensor type	Sensor ID	Description	Unit
Thruster	10	Azimuth angle feedback	Deg
	11	Azimuth angle order	Deg
	12	RPM feedback	%
	13	RPM order	%
	14	Running status	Bool
MRU	15	Ship heading	Deg
	16	Ship heave	m
	17	Ship pitch	Deg
	18	Ship roll	Deg
	19	Ship yaw rate	Deg/s
	20	Ship pitch rate	Deg/s
	21	Ship roll rate	Deg/s
	22	Ship vertical velocity	m/s

information on wind direction and speed, and wave related information such as wave height and wave period. The detailed environmental perception data can be found in Table 4.3.

There is a Palfinger crane onboard Gunnerus, as shown in Fig. 4.2. It is used for loading and unloading heavy objects, such as lifeboats and suction anchors.

The crane can reach out up to 20.5 m with a safe working load of around 2200 kg. Table 4.4 lists the specifications of the Palfinger crane. In order to grasp the crane operation situation, we have installed several types of sensors on it. The sensor location is illustrated in Fig. 4.3 and the corresponding sensor description is listed in

Table 4.3 Environmental perception data sensed onboard the R/V Gunnerus

Sensor type	Sensor ID	Description	Unit
GPS	23	Latitude	Deg
	24	Longitude	Deg
	25	Ship course relative to true north	Deg
	26	Surge velocity	Knots
	27	Sway velocity	Knots
	28	Speed over ground	Knots
Wind sensor	29	Wind direction	Deg
	30	Wind speed	m/s
Wave radar	31	Significant wave period	s
	32	Significant wave height	m
	33	Primary wave peak period	s
	34	Primary wave peak direction	Deg
	35	Total energy mean direction	Deg

Fig. 4.2 Palfinger crane onboard R/V Gunnerus

Table 4.4 Palfinger crane specification

Parameters	Value
Max lifting capability	15125.0 kg
Max lifting moment	610.0 kNm
Max hydraulic outreach	20.5 m
Max operation pressure	300.0 bar
Slewing torque	68.7 kNm
Pedestal diameter	834.0 mm
Deadweight	5960.0 kg

Table 4.5. A slewing sensor was installed inside the crane column, between the slip ring unit and the hydraulic swivel connector. Two pressure sensors were installed on the main cylinder, aiming to get the corresponding pressure in crane operation. Two angle sensors, one on the main boom and the other one on the outer boom, were installed to measure the tilt angle of crane booms. A length sensor was installed inside the cable drum on the outer boom of the crane, thus enabling the measure of crane elongation.

In summary, from Tables 4.1, 4.2, 4.3, and 4.5, there are in total 40 channels of sensor data that are utilized in the digital twin system.

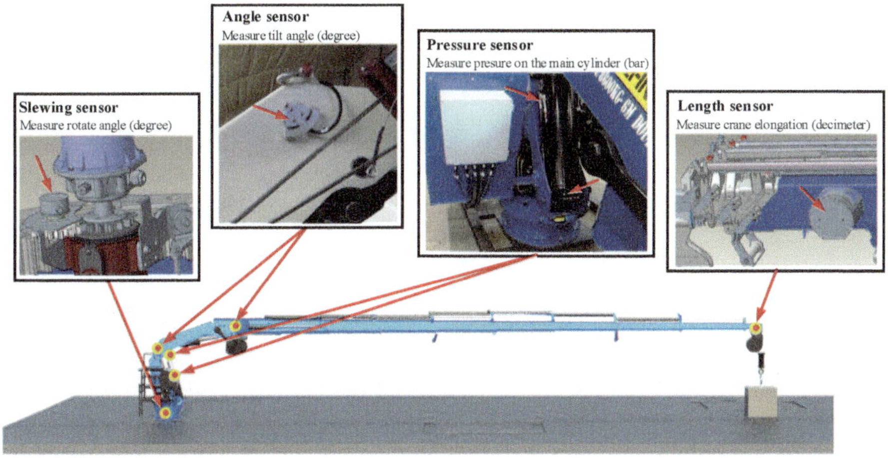

Fig. 4.3 Sensors deployed on Palfinger crane onboard Gunnerus

Table 4.5 Crane status data sensed on the Palfinger crane

Sensor ID	Description	Unit
36	Slewing angle of the base	Deg
37	Main cylinder pressure	Bar
38	Tilt angle of the main boom	Deg
39	Tilt angle of the outer boom	Deg
40	Extension length of the outer boom	Decimeter

4.2 Data Transmission and Storage

To develop a digital twin system, a huge amount of field data sensed by onboard sensors is needed. However, in most cases, the physical counterparts have lower computational power due to space and budget constraints. Therefore, it is challenging to directly develop the twin system onboard. Instead, establishing the twin system onshore would be an alternative solution, from where one can make use of high performance computers to model and train the twin system.

To this end, we build a remote control center, as shown in Fig. 4.4, for twin system implementation. Three control levels are considered, from monitoring, decision making, to emergency takeover. In the first level, we just need to monitor what is happening in the operation; whereas in the second level, supporting information could be generated at the center for a more comprehensive understanding of operation condition, not only at the present, but also in short future. If needed, the support information could be further forwarded to the onboard operator via the "instruction platform" shown in Fig. 4.4. The final level deals with contingencies, especially when the onboard operators take no action during an emergency. So far, the first two levels

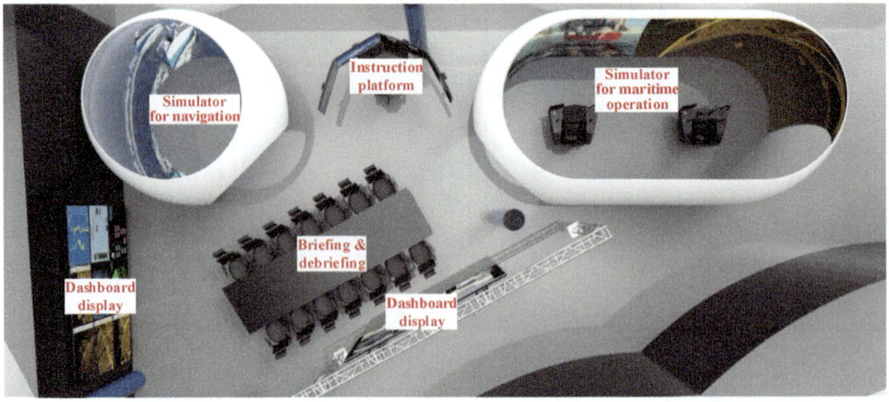

Fig. 4.4 Layout of the remote control center

have been achieved, including a dashboard system as seen in Sect. 4.3, and advanced functions of the digital twin system introduced in Chaps. 7, 8, and 9.

Sensor data, from the context, is the basis of the remote control center. In this project, we utilize the message queuing telemetry transport (MQTT) network protocol—a standard messaging protocol for the Internet of Things (IoT) for data transmission. The MQTT has a publish-subscribe architecture that contains four components, i.e., publisher, subscriber, broker, and topic. The topic could represent a concrete sensor channel, represented by a string consisting of one or more levels separated by a forward slash. In our case, sensor IDs in Tables 4.1, 4.2, 4.3, and 4.5 could be considered the topics of MQTT. The broker serves as a hub that routes and distributes all messages. The publisher is responsible for sending messages with topics to the broker, whereas the subscriber receives messages by subscribing to topics of interest from the broker.

Here, we briefly introduce how the onboard sensor data is transmitted ashore and stored into a database for modeling and training of the twin system. First, a next unit of computing (NUC) PC is deployed on Gunnerus. It performs as an MQTT publisher to publish data messages at a frequency of 1 Hz from different data sources on demand. Then, a local server at the remote control center is set up as an MQTT broker for forwarding received messages from the NUC PC. Next, a local PC is utilized as an MQTT subscriber. It not only receives MQTT messages, but also processes and stores sensor data into a database. After that, proper datasets are generated from the database for developing the twin system. Finally, the twin system is applied to analyzing real-time data from Gunnerus to enhance operational planning and decision-making. The scheme for Gunnerus data transmission and storage is illustrated in Fig. 4.5.

Regarding data storage, we use PostgresSQL—an advanced, enterprise-class, and open-source relational database system for data management, including insert, update, select and delete. Figure 4.6 illustrates an entity-relationship diagram, which represents how we add information including column name, column data type, and

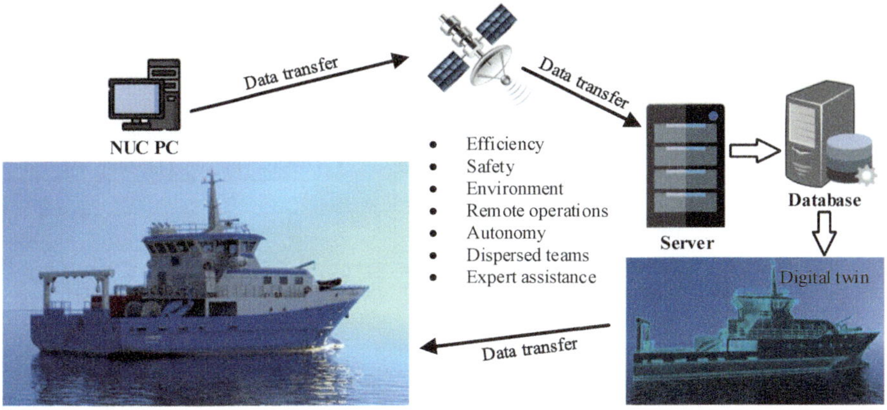

Fig. 4.5 Data transmission and storage scheme for digital twin system development

relationships between tables to the database. A sensor table is created, containing the 40 sensor IDs and their corresponding types and names. In addition, four categories of tables, corresponding to the sensor data from Palfinge crane, ship engine, ship motion, and environment, respectively, are incrementally generated. A threshold is set so that a new table can be created once its predecessor table's rows exceed that threshold. The purpose is to ensure that all tables in the four categories would not affect the efficiency of the query.

It is noted that in Fig. 4.6, the four categories of tables contain a column 'Event'. An event is a special user-defined condition associated with a table in the four categories. It can be defined based on either a single sensor source or a combination of multiple sensor sources. The event concept serves two purposes. First, it could facilitate the generation of datasets based on specific events for digital twin development. Second, if we delete sensor data that does not contain events, it will help to reduce the database size significantly. An event table is thus established, as shown in Fig. 4.6, to summarize the currently defined events. Table 4.6 lists some frequently occurring events. More events can be added into the table if required by any new applications.

4.3 Dashboard System Development

To find out what has happened, or is happening, or may happen to Gunnerus, it is necessary to find a way of displaying various types of sensor data in one place. A dashboard is such an information management system that can convey different, but related information in an easy-to-digest form. It is useful for monitoring, measuring, analyzing relevant data, and providing decision-making support in key areas. In our

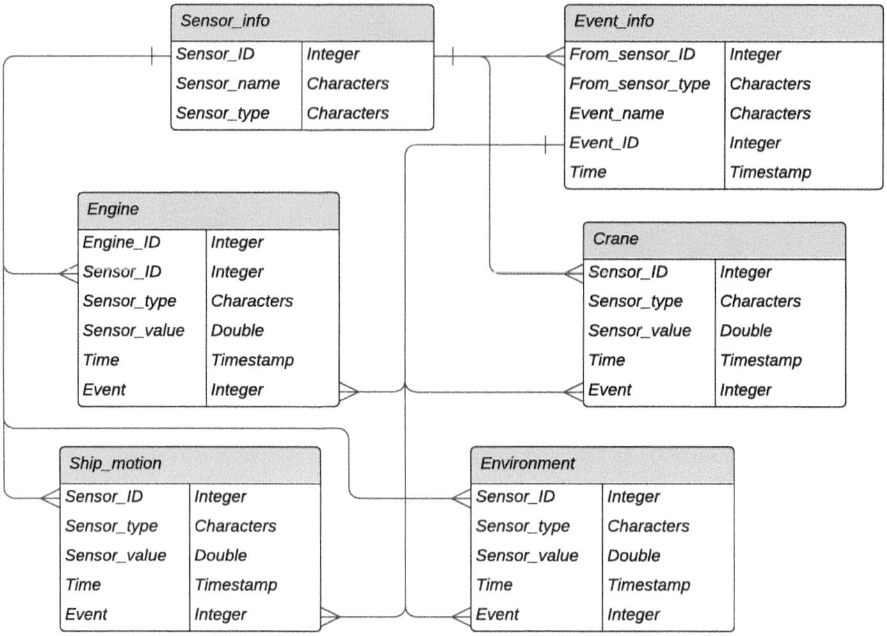

Fig. 4.6 Entity relationship diagram model in Gunnerus database

Table 4.6 Examples of Gunnerus event definition

Events	Sensor source(s)
Engine temperature alarm	Engine sensor
Engine pressure alarm	Engine sensor
Ship pose alarm	Ship motion sensor
Ship navigation	Ship motion sensor, environment sensor
Ship docking	Ship motion sensor, environment sensor
Ship dynamic positioning	Ship motion sensor, environment sensor
Crane in motion	Crane sensor
Crane lifting	Crane sensor, ship motion sensor
High sea state alarm	Environment sensor

case, considering the facts that both live sensor data via MQTT from Gunnerus and collected sensor data in database from the remote control center are available, a dashboard system is designed to translate these data into easily understood visuals, such as charts, graphs, and 3D animation, enabling crews at the remote control center to stay informed about current/historical situation at a glance.

Data visualization especially environmental mapping is the most important element for dashboard development, as it is the fundamentals for representing ship operation status in RCC. There are multiple ways the mapping process can be per-

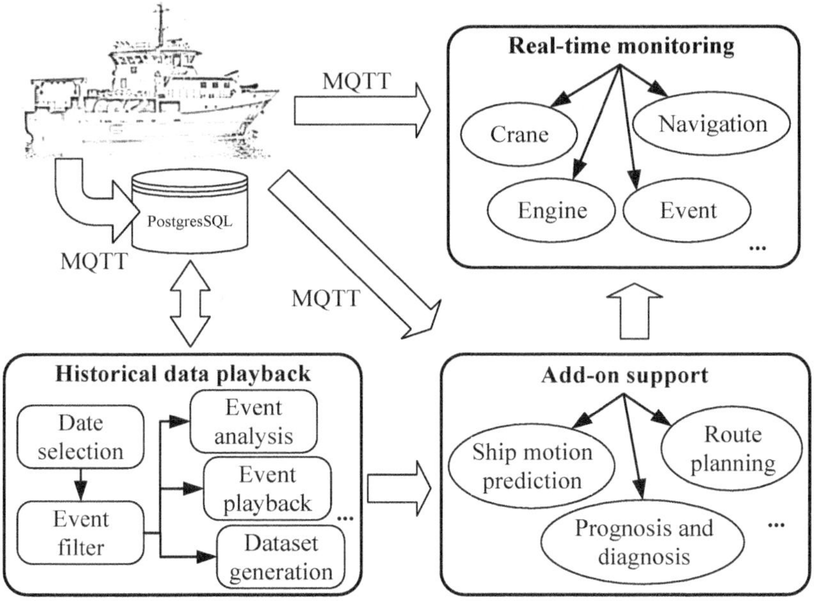

Fig. 4.7 Gunnerus dashboard system

formed and what kind of presentation of the world is created depends on the application, where the maps are needed and what sensors are used for perceiving the environment. Here, Mapbox[1] is utilized for ship location positioning and environment visualization. Assets such as 3D models of the ship engine, the Gunnerus hull, the components of the Palfinger crane are built and converted to the format compatible with the ThreeJS loader, from which these 3D models can be added into Mapbox. Moreover, Javascript is used at the backend for logical processing with human-computer interaction, such as data query, data visualization, and ship-crane assembly and animation.

Figure 4.7 depicts the functional scope of the dashboard system. Taking advantage of MQTT, 1 Hz sensor data flow could be fed into the dashboard and the states of Gunnerus including the engine, the crane and the ship motion itself could be presented in the form of animation and dynamic charts in real-time. This means, onshore crews in RCC could monitor marine operations by locating the Gunnerus vessel and manipulating the 3D visualization from any angle. In addition, if any of the predefined events as described in Sect. 4.2 is identified, the event name and the corresponding start time will be displayed in the monitoring interface.

To find out what has happened to the Gunnerus vessel, a "historical data playback" function has been implemented in the dashboard. It is based on an event-driven concept that data are categorized by date and different types of events. Simple statistical

[1] https://www.mapbox.com/.

analysis on certain types of events is provided, such as total fuel consumption, total operation time and distance traveled. For any event of interest, one can replay that specific event in animated form as fast as in real-time or even faster. It is noted that the "historical data playback" function also provides an interface for event collection and dataset generation, which facilitates digital twin modeling and training.

The last highlight of the dashboard is the "add-on support" function. The word "add-on" here indicates a means of integrating advanced tools into the dashboard. From our previous work in [119], a digital twin of Gunnerus was built, in which advanced tools such as ship motion predictor, auto-docking controller and engine anomaly detector are available and thus can be added into the dashboard for decision support of marine operation. As a result, the dashboard could make use of the tools from the twin system and display the support information to the front end of "real-time monitoring".

4.4 Dashboard Case Studies

In this section, we present three case studies to verify the effectiveness of the dashboard deployed in our RCC at NTNU Ålesund, Norway.

4.4.1 Real-Time Ship-Crane Operation

A field experiment of Gunnerus was conducted at Ålesund, Norway on April, 2022. The task is to perform dynamic positioning of Gunnerus while manipulating the onboard Palfinger crane to lift a lifeboat up and down around the ship. In the meantime, we run the real-time monitoring in the dashboard at the RCC. The experiment lasted about 30 min and a drone video of the field operation was recorded, with the purpose of post-analysis and comparison with the animation displayed at the remote control center.

Figure 4.8 shows the comparison between the field operation and the animation.[2] They are video screenshots at time 72, 402, 643, 851, 1142 and 1340 s, corresponding to the moments when hooking up the lifeboat, rotating toward ship stern, elongating the crane on starboard, lifting and lowering the lifeboat at ship stern, and shorten the crane on the port side of the ship, respectively. Due to the battery limit of drone video, there was no live video after around 20 min of crane operation. That is why the last view at the 22nd minute of the crane operation in Fig. 4.8 shows only the animation. Despite this, the rest of animation is almost consistent with the real operation through our observation. It could be seen that the movement in the animation is close to what happened in reality. Due to the low frequency of data transfer, the animation is not

[2] https://org.ntnu.no/intelligentsystemslab/gunnerus/dt_realtime_crane_monitor.html.

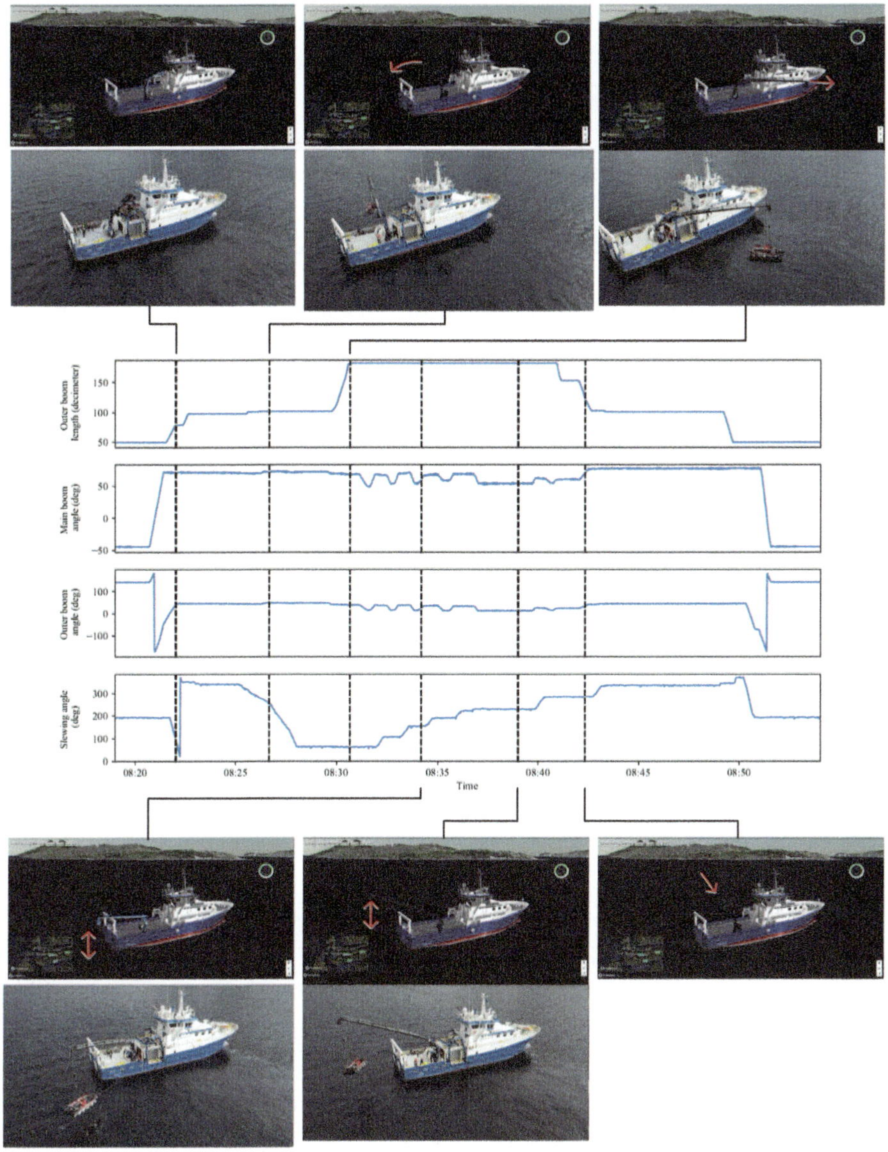

Fig. 4.8 Real-time monitoring of ship-crane operation on Gunnerus

smooth enough, but potential improvements can be achieved in future by interpolating sensor data.

4.4.2 Ship Docking Replay

From our previous work [156], we have set proper thresholds on several variables, such as ship speed over ground, fuel consumption, and thrust RPM order for defining ship docking events. The data from the docking moment back to 8–10 min prior to docking is considered the corresponding data segment for representing a complete docking operation. Through the "historical data playback" function, how the Gunnerus ship approaches a dock could be visualized for analysis or evaluation.

Figure 4.9 is a replay example of ship docking at Ålesund port. The replay interface not only provides a drag-and-drop time bar, but also supports fast replay, allowing animation playback up to 10 times faster than real-time. In addition, the dynamic charts aside present the status changes of Gunnerus ship over time, facilitating remote operators to fully understand ship behavior. For example, from sub-figures ① to ③ of Fig. 4.9, Gunnerus maintains a speed about 11.7 m/s with a fuel consumption around 97 liter/min. It is obvious to see the drop down of fuel consumption and the decrease of surge speed in ⑤ of Fig. 4.9 when Gunnerus approaching the port. Sub-figures ⑥ to ⑧ of Fig. 4.9 illustrates how Gunnerus turns toward the dock at a relatively low velocity. It is considered that further analysis and comparison of docking behaviors via the replay interface can be implemented in the future if the data segments of multiple docking events at the same port are detected. In addition, if enough docking samples are collected, advanced tools such as docking motion prediction and automatic docking control could be generated in the digital twin system.

4.4.3 Ship Motion Prediction

Ship motion prediction is considered as a way of decision support for ship navigation. There are plenty of methods that can achieve the goal, including data-driven methods [157], model-based methods [158], and their combinations [159]. To verify the "add-on support" function of the Gunnerus dashboard, a simple kinematic model-based predictor is applied to "real-time monitoring" interface.

The predictor running in our twin platform [119] takes current ship position, heading, surge and sway velocities and yaw rate as inputs, and produces future ship position in 20 s ahead. Figure 4.10 illustrates the prediction results for Gunnerus ship in navigation at a speed around 11.5 m/s. It is noted that in sub-figures ①, ③ to ⑦ of Fig. 4.10, Gunnerus is cruising forward and its surge speed dominates the motion, resulting in a straight line prediction; whereas in ② and ⑧ of Fig. 4.10, Gunnerus steers but maintains the cruise speed, thus both the surge speed and the yaw rate play an important role in generating a turning motion prediction.

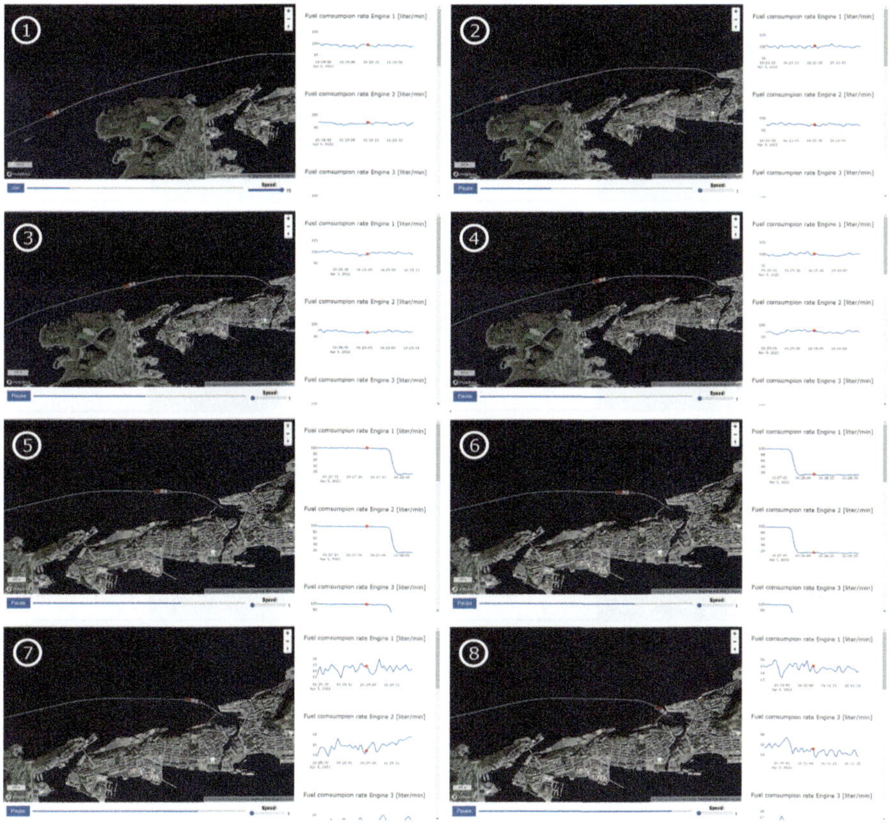

Fig. 4.9 Ship docking replay at Ålesund port

Since the computational cost of the predictor is low, real-time predictions can be achieved. The prediction performance may be relatively low compared to other prediction methods but acceptable to onshore crews in RCC, as the target is to reveal navigation intention rather than high precision of ship trajectory in short future.

4.5 Summary

In this chapter, we provide an overview of the data infrastructure of R/V Gunnerus. Sensor data from ship engine, ship motion, ship crane and environmental perception are collected and transmitted from an onboard NUC PC via a 4G connection to a local server in our remote control center. The data are further stored into a database for digital twin development. A dashboard system is developed, which enables real-time monitoring of Gunnerus together with advanced decision support and historical

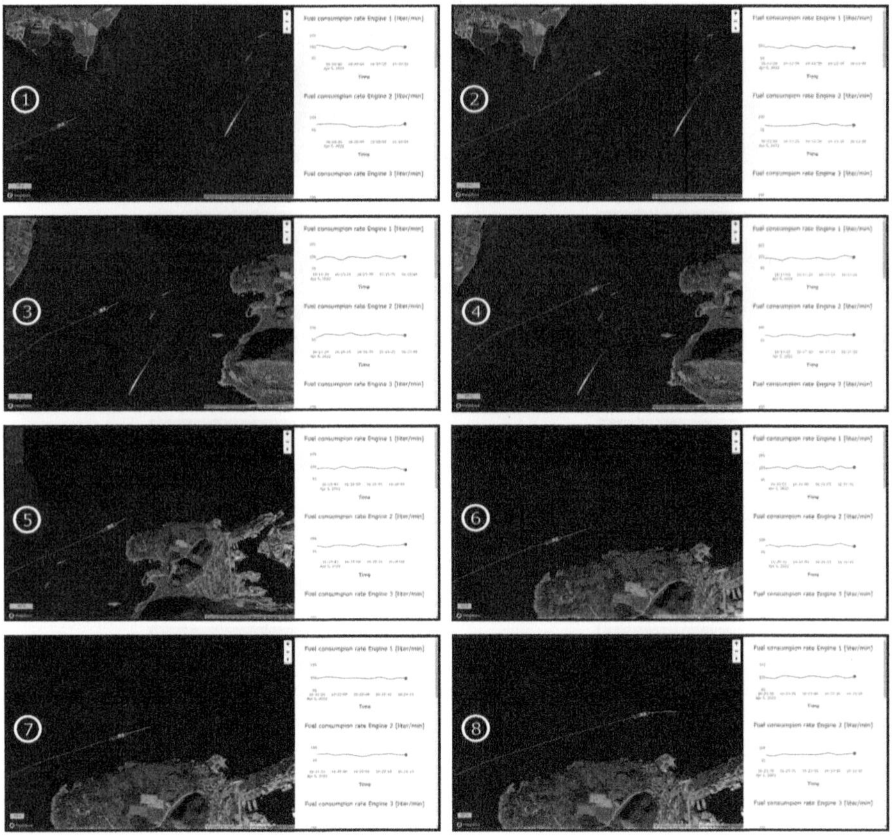

Fig. 4.10 Ship motion prediction during Gunnerus voyage

replay. Through three case studies, we verify the effectiveness of the IoT solution for data collection, transmission and storage.

Guoyuan Li received the Ph.D. degree in computer science from the Department of Informatics, Institute of Technical Aspects of Multimodal Systems, University of Hamburg, Hamburg, Germany, in 2013. In 2014, he joined the Department of Ocean Operations and Civil Engineering, Intelligent Systems Laboratory, Norwegian University of Science and Technology (NTNU), Ålesund, Norway, where he is currently a Professor in ship intelligence. He has published more than 150 articles in the areas of his research interests, including the modeling and simulation of ship motion, autonomous navigation, intelligent control, optimization algorithms, and the locomotion control of bioinspired robots.

Finn Tore Holmeset has a background as a Marine Chief Engineer, with more than 13 years seagoing experience onboard different ship and holds a certificate at the highest level (M1). In 2014 he joined the Department of Ocean Operations and Civil Engineering, at Norwegian University of Science and Technology (NTNU), Ålesund, where he received his master's degree in management of Demanding Marine Operations in 2018. He is currently undertaking a Ph.D. in maritime technology and its impact on marine engineers.

Houxiang Zhang received the Ph.D. degree in mechanical and electronic engineering from Beihang University in 2003 and the Habilitation in informatics from the University of Hamburg, Germany in 2011. He is currently a full Professor with the Department of Ocean Operations and Civil Engineering, Faculty of Engineering, Norwegian University of Science and Technology (NTNU), Norway. From 2011 to 2016, he also hold a Norwegian National GIFT Professorship on product and system design funded by the Norwegian Maritime Centre of Expertise. He has engaged into two main research areas, including control, optimization and AI application especially on autonomous vehicle; and marine automation, digitalization and ship intelligence. He has applied for and coordinated more than 30 projects supported by Norwegian Research Council (NFR), German Research Council (DFG), EU, and industry. In these areas, he has published over 300 journals and conference papers as the author or coauthor. He was nominated and selected as a member of the Norwegian Academy of Technological Sciences in 2019, and a member of Academy of the Royal Norwegian Society of Sciences and Letters in 2024.

Hans Petter Hildre is professor and head of the Department of Ocean Operations and Civil Engineering at the Norwegian University of Science and Technology (NTNU). His area of interest is product design and system architecture design. Hans Petter is the director for Centre for Research Driven Innovation (SFI-MOVE) within marine operations. This is cooperation between NTNU, SINTEF, University Sao Paulo and 15 companies at the west coast of Norway. Professor Hildre is head of research in national program Global Centre of Expertise Blue Maritime, project leaders in several research projects, member of the board in 5 companies, and has a number of patents.

Chapter 5
Models and Sub-domain Models for Digital Twins

Yingguang Chu, Guoyuan Li, Vilmar Æsøy, and Houxiang Zhang

Hybrid modeling, physics-based (It is considered the equivalent of model-based, and we continue to use this terminology in this chapter unless otherwise stated.) and data-driven modeling, co-simulation, and digital twin development are all essential components of digital twin development for multi-domain operation systems (MDOS). This chapter elaborates on applications of hybrid modeling, co-simulation, and digital twin technology to MDOS modeling. It discusses the advantages of using these technologies to improve the performance, reliability, and scalability of MDOS modeling. It also discusses the challenges associated with implementing these technologies and the potential solutions to address them. An overview of existing research and applications is given. Finally, model development towards the applications and implementations for the Gunnerus digital twin ship is described by using its reference models.

5.1 Physics-Based and Data-Driven Modeling

Physics-based and data-driven modeling are two main approaches to simulations of MDOS. Physics-based simulation involves creating a mathematical model of the system being simulated. This technique is used to accurately simulate the behavior of a system. The models are based on the underlying physical laws and relationships that govern the system. The simulation of the models generates approximations of the behavior of the system over time. This approach is particularly useful when the underlying physical laws are well understood and can be accurately represented mathematically. It can be particularly useful for design and optimization, as it allows for the identification of the most important features that influence the system's behavior. However, physics-based simulation can be limited by the accuracy of the mathematical models. If the models are not accurate, the approximations generated by the simulation are also inaccurate. In addition, physics-based simula-

H. Zhang et al., *Digital Twins for Vessel Life Cycle Service*,
https://doi.org/10.1007/978-981-97-4768-9_5

tion can be time-consuming and resource-intensive, as it requires the development and implementation of a mathematical model, especially, for complex multi-domain systems.

Data-driven simulation, on the other hand, relies on empirical data to generate approximations and predictions about the behavior of the system being simulated. This technique is often used when the behavior of a system is difficult to model using traditional physics-based methods. The simulation model is based on data obtained from measurements or experiments, and the model is refined and updated as new data becomes available. This approach is useful when the underlying physical laws are not well understood or when the system being simulated is too complex to be accurately modeled mathematically. Data-driven simulation is often faster and more flexible than physics-based simulation. This makes data-based simulation particularly useful for applications where real-time predictions are needed. However, data-driven simulation can also be limited by the quality and availability of the data used to generate the simulation model. If the data is not representative of the system being simulated, or if it is not available in sufficient quantities, the predictions generated by the simulation will be compromised.

Both physics-based and data-driven simulations are important tools for understanding and predicting the behavior of complex systems. The choice between these two approaches often depends on the specific requirements of the problem, including the accuracy and speed of the simulation, the availability of data, and the underlying physical laws that govern the system.

5.2 Hybrid Modeling

Hybrid modeling combines different modeling approaches to represent the behavior of dynamic systems. This approach has become increasingly widely applied for the analysis and optimization of dynamic operation systems, which typically involve complex interfacing between physical domains, communications, and interactions between hardware, simulation models, and human-in-the-loop.

One of the key advantages of hybrid modeling is its ability to capture the complex interactions between different processes in dynamic operation systems. This provides a more comprehensive representation of the system's behavior, allowing for a better understanding of the system's dynamics and identification of key factors that influence its behavior. A major advancement in hybrid modeling is the integration of data-driven and physics-based approaches. This integration allows for the creation of hybrid models that combine data-driven models, which are based on empirical data, with model-driven models, which are based on physical laws and relationships. The combination of these two approaches provides a more accurate representation of the system behavior, taking into account the strengths and weaknesses of both approaches.

As an example, [160] proposes a physics-data cooperative ship dynamic model for the docking operation; a knowledge-based ship model serves as a physics foundation with supportive data-driven models compensating a single-step-ahead velocity prediction error made by a physics foundation model. Neural networks trained with onboard sensor data are employed in supportive data-driven models. Reference [11] propose a physics-data cooperative modeling approach based on a rough ship numerical model and a few operational data to enhance the model quality. The prior knowledge leveraged by the ship's numerical model is integrated into the neural network as informative inputs, and the informed neural network calibrates the bias between model outcomes and actual states in principle.

As a short summary, hybrid modeling has made significant advancements in recent years. It provides the following features including the integration of data-driven and model-driven approaches, cloud-based simulation platforms, integration with optimization techniques, and the integration of machine learning and artificial intelligence. These developments have made hybrid modeling an essential tool for the analysis and optimization of dynamic operation systems.

5.3 Object-Oriented Modeling

Regarding domain specific system modeling, object-oriented modeling (OOM) is a widely used approach for complex MODS. Sections 5.3 and 5.4 describe the application of Bong Graph (BG) as a special form of OOM for MDOS modeling [55]. OOM originated from software model development and resulted in the establishment of the Unified Modeling Language (UML) in the 1990s [34]. Later, SysML as an extension of UML, is a graphical modeling language that provides a general purpose modeling language to support specification, analysis, design and verification of dynamic systems [35]. Known as the four pillars of SysML: the requirements, the behavior, the structure and the parametric, in total nine diagrams are used to reflects the various aspects of a system model. For digital twin development of MODS, OOM provides an effective architecture to represent the physical system being modeled, whether it is physics-based or data-driven.

The architecture of a component physics-based on the OOM paradigm for simulation is shown in Fig. 5.1. The behavior of an object is separated from its structure. Specifically, the structure model defines the representation and properties, including the parameters and user interfaces for model manipulation. The behavior model describes the dynamics of the physical system and the simulation interfaces for model interaction. This includes the parameters, state variables, input and output of the component model to others where any physical connection exists. Model classification is established based on the different complexity levels of modeling in describing the behavioral characteristics of the system. For one component model, several implementations share the same set of input and output to its outside world, but can contain different parameters and state variables used in describing the system's dynamic behavior.

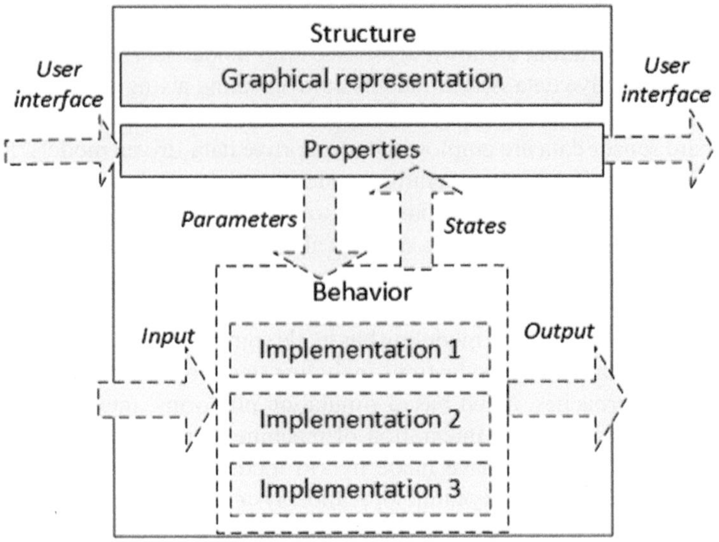

Fig. 5.1 Model architecture of an object-oriented component model

The OOM process, first and foremost, involves the identification of objects, their properties, their functions, and their relationships. An object can be anything that represents a physical entity in the system being modeled. They can be parts, machines, sensors, or even human operators. Each object has properties that describe its characteristics, such as its size, weight, state, and location. Objects also have relationships with other objects in the system, which define how they interact with each other.

Once the objects and their properties are defined, the relationships between the objects can be established. These relationships can be classified as either structural or behavioral. Structural relationships define how objects are connected, for example, which parts are connected to which machines. Behavioral relationships define how objects interact with each other, for example, how a sensor detects a temperature change and triggers a response.

OOM also allows for the creation of classes, which are templates for objects. Classes define the properties and behaviors that objects will have. For example, a class for a machine might include properties such as its size, weight, and power consumption, as well as behaviors such as starting up, shutting down, and producing products. By creating classes, engineers can streamline the modeling process and reduce the amount of redundant code required.

OOM provides a convenient approach to modeling for digital twin system development. By using this approach, engineers can create an accurate and detailed representation of the physical system being modeled, including its components, properties, and relationships. This enables the system to be simulated and optimized, which can save time and money in the development process.

5.4 Bond Graph as a Special Form of OOM for MDOS

Before the advent of OOM and Modelica, the Bond Graph (BG) method was one of the most effective approaches to MDOS modeling. It describes the physical system by identifying its energetic structure and represents the energy flow using generalized variables: the effort e, the flow f, the displacement q and the momentum p. For example, in mechanical domain, these variables are force F or torque T, velocity v or angular velocity $\dot{\theta}$, the displacement s or angle $theta$ and the momentum P or angular momentum L. Correspondingly, in the hydraulic domain, these variables represent the pressure P, the volumetric flow $\dot{V}$, the volume V, and the pressure momentum p_p. As shown in Fig. 5.2, the so-called tetrahedron of state depicts the interrelations of the three basic single-port BG elements, including the inertia $\boldsymbol{I}$, the capacitor $\boldsymbol{C}$ and the resistor $\boldsymbol{R}$ [161]. Multi-ports BG elements, special types of BG fields and equation-based iconic diagrams for specific applications can be created following the same principles. These BG elements and iconic diagrams provide generalized forms for representations of the component models in different energy domains.

BG modeling can be seen as a special form of OOM modeling [162]. A Modelica BG library, in effect a translation of BG elements in Modelica has been presented by Broenink and Cellier [163, 164]. Later, an export filter for BG models implemented in a software tool called 20-sim was developed to generate Modelica code [165]. The generation of Modelica code from the existing basic BG elements and block diagram appears quite straightforward. As stated by Broenink, since BG method existed before OOM was applied to physical system modeling, BG can be seen as a form of OOM approach as it shares the key features of OOM paradigm as follows:

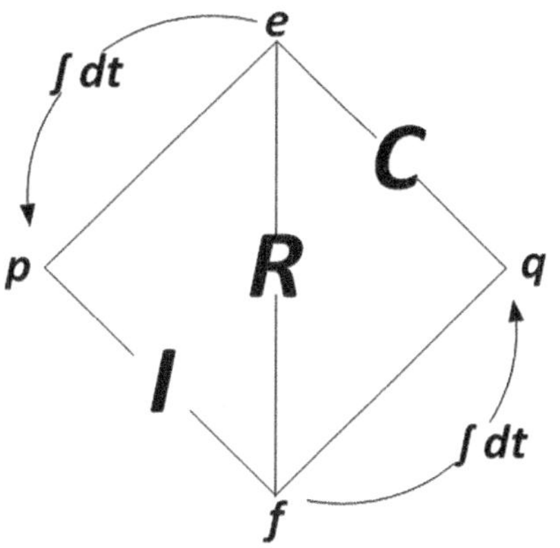

Fig. 5.2 The tetrahedron of state

1. A-causal modeling: Computation causality is not needed in modeling, but needs to be decided in compiling the model into computable code. A-causal modeling means that the equations are written as true equations that state the mathematical principles rather than determining the actual computational sequence. In this way, the interfaces of the sub-model can be defined by variables that are not committed to any role for computation. This is essential if the internals of the sub-models are to be completely encapsulated.
2. Encapsulation: The interaction with a model is only accessible via well-defined interfaces with its "outside" world. The use of the model is not constrained by the internal specifications. This allows for the reuse of the model or certain parts of model while keeping the rest untouched.
3. Hierarchy: A physical system is hierarchical in nature and can be decomposed into sub-systems. Thus, the model of a system is composed of sub-models that may contain lower levels of sub-models as well.
4. Inheritance: Based on a hierarchical relationship, the properties of a sub-model are inherited by its upper class of sub-models and shared by the lower class of sub-models.

Using some examples in the case studies, the following subsections describe the construction of thier behavior models with different complexities based on OOM and BG.

5.4.1 Behavior Model of the Hydraulic Cylinder Implementation 1—Simplified

In the demo case, ship-crane operation, the Palfinger crane on the Gunnerus ship is powered by hydraulic actuation systems. The hydraulic cylinder converts hydraulic power into mechanical power through the pressurized liquid in the cylinder chambers and the piston-rod connected to the crane boom via a constraint. Due to the inherent nature of fluid dynamics that pressurized liquids are nearly in-compressible, solving the dynamic equations requires very small time steps, hence a long simulation time. For real-time simulation of crane operations, a simplified behavior model of the hydraulic system is usually preferred. Assuming that the liquid is incompressible, the behavior of the hydraulic cylinder can be simply represented by the conversion of the hydraulic power to the mechanical translation power. The first BG model implementation representing the simplified hydraulic cylinder is shown in Fig. 5.3.

According to the preferred causality, Eq. 5.1 holds for the pressure at the piston side P_a. The pressure at the rod side P_b is given by the return line pressure represented by an effort source.

$$P_a \cdot A_a = P_b \cdot A_b + F \tag{5.1}$$

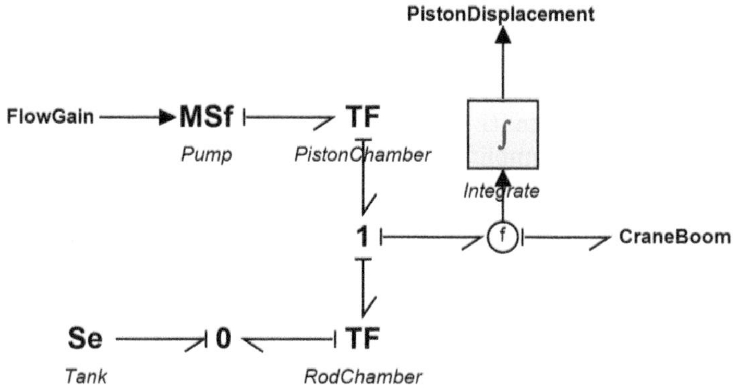

Fig. 5.3 Simplified behavior model of the hydraulic cylinder represented by Bond Graphs

where A_a, A_b denotes the area of the piston at either side, F is the external force acting on the cylinder piston.

The modulated flow source represents the liquid flow from the pump into the cylinder. The volumetric flow rate is given by Eq. 5.2.

$$\dot{V}_a = f \cdot \forall \cdot \omega_{cyl} \tag{5.2}$$

where f is the flow gain of the pump flow, $\forall$ is the displacement of the pump per revolution, and ω_{cyl} is the speed of the pump in revolution per second.

Two transformers represent the power conversion at both sides of the piston. Since the liquid is incompressible, the velocity of the cylinder piston is given by Eq. 5.3.

$$v = \frac{\dot{V}_a}{A_a} \tag{5.3}$$

Accordingly, the volumetric flow rate at the tank side of the cylinder piston can be derived by Eq. 5.4.

$$\dot{V}_b = v \cdot A_b \tag{5.4}$$

5.4.2 *Behavior Model of the Hydraulic Cylinder Implementation 2—Extended*

The simplified model above describes simply the motion of the hydraulic cylinder. Other properties include the pressures and flow rates at both chambers, valve characteristics are not modelled. These can be established in the more extended model, including the flow compressibility, inertia of the flow flux, friction loss through the pipeline, pressure drops over the valves and flow restrictions. The extended model describes the behavior of all components as shown in Fig. 5.4, including the hydraulic power unit (HPU), pressure control valve (PCV), direction control valve (DCV), counter-balance valve (CBV), and the cylinder. The BG implementation of the extended hydraulic cylinder model is shown in Fig. 5.5. The sub-model shares the same interfaces as the simplified model implementation. In the simplified model, the flow gain is applied directly to the flow source representing the pump flow. In the extended model, valve control is used. The flow to the cylinder is controlled by a four-way-three-position DCV, a pressure compensator for the DCV and a CBV at the supply line when the cylinder is exposed to negative loads (when the piston velocity and loading force have the same direction). The constitutive equations describing the behavior of these components are presented as follows.

The power supply of the HPU is defined as a pressure compensated displacement pump. The flow rate from the pump is given by Eq. 5.5.

$$\dot{V} = \frac{P_{set} - P}{\Delta P} \cdot \dot{V}_{max} = \frac{P_{max} - P}{\Delta P} \cdot \forall \cdot \omega_{cyl} \tag{5.5}$$

where P_{set} is the set point pressure of the pump, ΔP is the pressure deviation from set point for full pump flow $\dot{V}_{max}$.

The pipe flow represents a lumped model of the fluid dynamics through the pipeline including the compressibility of the liquid, the inertia of the flow flux and the

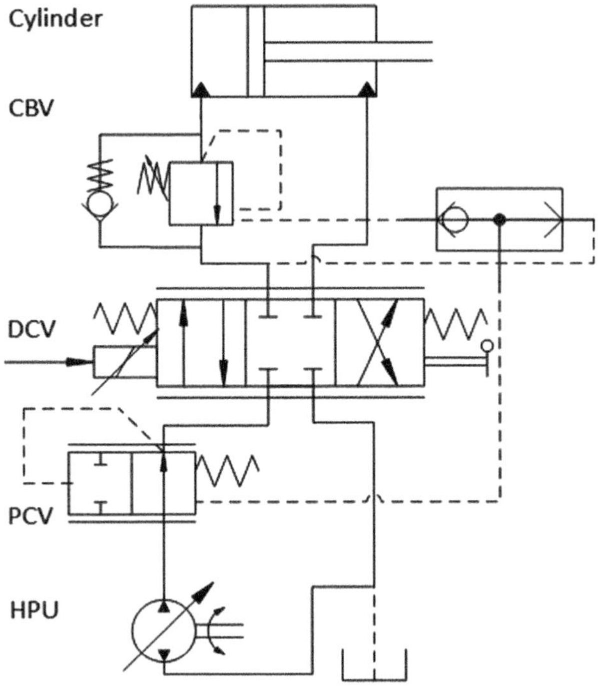

Fig. 5.4 Simplified hydraulic diagram of the Palfinger crane

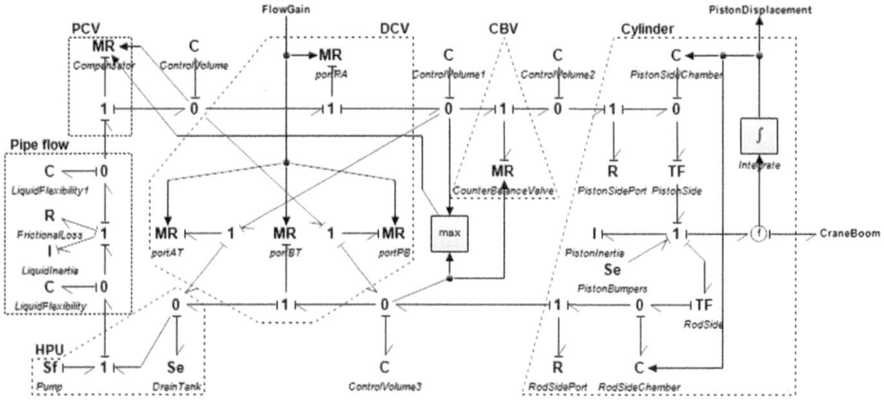

Fig. 5.5 Extended behavior model of the hydraulic cylinder implemented with Bond Graphs

friction loss. The control volumes represent the compression of the small volumes of liquid between the hydraulic components.

The compressibility of the liquid is given by Eq. 5.6.

$$\Delta P = \frac{B}{V} \cdot \Delta V \tag{5.6}$$

where B is defined as the bulk modulus of the liquid and considered as constant, V denotes the original volume of the liquid which can be the physical volume of the pipe segment or the control volumes, ΔV represents a small decrease in the volume of the liquid.

The expressions for the compressibility of the liquid in the cylinder chambers are slightly different because the original volumes of the liquid change when the cylinder piston moves. The pressures in chambers A and B are given by Eqs. 5.7 and 5.8 respectively.

$$\Delta P_a = \frac{B}{V_{a0} + A_a \cdot (x_0 + x)} \cdot \Delta V_a \tag{5.7}$$

$$\Delta P_b = \frac{B}{V_{b0} + A_b \cdot (s + x_0 - x)} \cdot \Delta V_b \tag{5.8}$$

where V_{a0} and V_{b0} denote the dead volumes of the cylinder chambers, x_0 and s denote the initial displacement and stroke of the cylinder piston.

The inertia effects of the liquid flow through the pipe are analogous to the mass for a body in translation. Assuming that the flux of the liquid flow in the pipe moves as a rigid body, the Newton's law becomes as written in Eq. 5.9.

$$\Delta P \cdot A = (\rho \cdot A \cdot l) \cdot \frac{\ddot{V}}{A} \tag{5.9}$$

where ρ is the density of the liquid, A is the effective section area of the pipe, l is the length of the pipe segment, $\dot{V}$ represents the volumetric flow rate through the pipe, hence, the velocity of the liquid slug becomes $\dot{V}/A$, and its acceleration becomes $\ddot{V}/A$.

The pressure momentum is defined as given by Eq. 5.10. As can be seen, the inertia coefficient is proportional to the mass density and length of the fluid flux, but inversely proportionate to the area.

$$p_P = \int (\Delta P \cdot A) dt = \frac{\rho \cdot l}{A} \cdot \dot{V} \tag{5.10}$$

The friction loss of the flow through a circular pipe is implied by a pressure drop related to the volume flow rate as written in Eq. 5.11.

$$\Delta P = \lambda \cdot \frac{l}{d} \cdot \frac{\rho}{2} \cdot (\frac{\dot{V}}{A})^2 \tag{5.11}$$

where λ is the friction factor of the liquid flow in the pipe and d is the diameter of the pipe.

The friction factor λ can be written as in Eq. 5.12 depending on the flow pattern of the pipe flow, whether it's laminar flow or turbulent flow.

$$\lambda = \begin{cases} \frac{64}{Re} & Re < 2000 \\ \frac{0.25}{(\log(\epsilon/(3.7 \cdot d) + 5.74/Re^{0.9})^2} & Re > 2000 \end{cases} \tag{5.12}$$

where Re is the Reynold's number, ϵ is the absolute roughness of the pipe.

The liquid flow through any valve or port causes a pressure drop in the direction of the flow. The relationship between the pressure drop and the volumetric flow rate can be established following approximately the nozzle equation, as given by Eq. 5.13. The direction of the flow is implied by the sign of the pressure drop ΔP.

$$\dot{V} = C_d \cdot A \cdot \sqrt{\frac{2 \cdot \Delta P}{\rho}} \tag{5.13}$$

where C_d is the discharge coefficient of the nozzle, A denotes the open area of the nozzle.

As shown in Fig. 5.5, the PCV, DCV and CBV are represented by modulated resistors. The opening area of the valves is controlled by modulated signals. more specifically, the PCV opens when the pressure difference across the DCV is bigger than the compensation pressure. The opening of the four ports of the DCV are controlled by the flow gain from the PI-controller, which determines the opening area of each port. The flow through the CBV is dependent on the pressures at both sides of the cylinder piston. The CBV is fully open when hoisting the load, and partly open when lowering the load in case of loose control of the load and cause negative pressure in the cylinder.

The velocity of the cylinder piston can be calculated by its momentum according to Newton's second law.

$$p = \int F dt = m \cdot v \tag{5.14}$$

where m is the mass of the cylinder piston.

The effort source in the cylinder model represents the bumpers of the piston, which generates the reaction forces upon contact with the piston. The reaction force of the bumper is given by Eq. 5.15.

$$F = \begin{cases} -k \cdot x - c \cdot \dot{x} & x <= -\delta x \\ 0 & 0 < x < s \\ -k \cdot (x - s) - c \cdot \dot{x} & x >= s \end{cases} \tag{5.15}$$

where k and c are the spring stiffness and critical damping factor of the bumper. x and s are the displacement and stroke of the piston.

5.5 Orchestra the Simulation Enabled by Co-simulation

Once the objects, properties, and relationships have been defined, the system can be simulated in a digital twin platform as described in Chap. 3. This platform provides a virtual environment where the modeled system can be tested and optimized. By simulating the system, engineers can identify potential problems and make changes to the design before the physical system is built. This can save significant time and cost by reducing the number of physical prototypes required.

Co-simulation is a simulation technique that allows for the integration of multiple simulation models to represent the behavior of complex systems. This approach is used to represent the interactions between different components of a system, providing a more comprehensive representation of the system behavior. In Chap. 3, co-simulation as a technique to combine multiple models into a single simulation is described, particularly, the development of the Open Simulation Platform (OSP) for the digital twin ship development.

Co-simulation is particularly useful for the development of MDOS to combine physics-based and data-driven models into a single simulation. By combining the two, a more accurate representation of the system can be created. One of the key advantages of co-simulation is its ability to capture the complex interactions between different components of a system. For example, a co-simulation can be used to represent the interactions between a power plant and the electrical grid, allowing for the evaluation of the impact of changes in the power plant on the grid. This provides a more accurate representation of the system behavior, allowing for a better understanding of the system dynamics and the identification of key factors that influence its behavior.

An important aspect of the orchestration of simulation models is the standardization of their interfaces. In many applications there is the need to design, simulate and execute a network of components from different users. A major advancement in co-simulation is the integration of different simulation models and tools. This integration allows for the creation of co-simulation models that combine models from different domains, such as electrical, mechanical, and thermal. The combination of these models provides a more comprehensive representation of the system behavior, taking into account the strengths and weaknesses of each model.

Since the last decade, the Modelica language has drawn the attention of both academic and industrial users. Modelica is a non-proprietary, object-oriented, equation-based language designed for the convenient modeling of complex physical systems. Basic models of physical components are defined in a declarative manner by their constitutive equations and connected with the outside world without implied causality. This makes the description of the physical systems more flexible and understandable than using causal or block-oriented modeling languages. It is also easier to use than scripting simulation code using procedural languages like C or FORTRAN [166]. Complex models can be built by connecting the basic models where the physical connections exist. Since the models are written in terms of generic differential-algebraic equations, components that define interactions between different physical domains can be combined without any restriction. In this regard, the SSP is an extension of the FMI standard to define complete systems consisting of one or more FMUs including its parameterization that can be transferred between simulation tools.

The INTO-CPS project has developed INTO-SysML models for co-simulation of a cyber-physical system [36]. The INTO-SysML profile customizes SysML for architectural modeling in a setting of multi-modeling and FMI co-simulation. It introduced specializations of SysML blocks to represent different types of CPS components, constituting the building blocks that enable a hierarchical description of the CPS architecture. SysML is a powerful, but formidable, technical language to master, especially for the designers and system engineers [37]. Its extensions, such as INTO-SysML, could be a better alternative for modeling if matured. Up to date, the application of OOM within dynamic operation systems is still at the initiative stage. The purpose of introducing the OOM method for dynamic operation systems is to facilitate the modularization and integration of model development. Thus, all different users can form an understanding on the same matter (model) and contribute their expertise on one sub-system without worrying about the details of others.

For the same purpose, the OSP project outlined the OSP-IS specification for maritime systems [148]. The standard and specification are developed aiming to enable faster construction of co-simulation simulators by simplifying the model connection process, specifically, for maritime system models.

The use of co-simulation has also been facilitated by the development of cloud-based simulation platforms. These platforms provide access to high-performance computing resources, allowing for the simulation of complex and large-scale systems. This has made co-simulation more accessible and cost-effective, enabling smaller organizations to take advantage of these technologies. In addition, the integration of co-simulation with optimization techniques has opened up new opportunities for the optimization of complex systems. Co-simulation can be used to identify the optimal operating conditions for a system, taking into account multiple objectives and constraints. This has made it possible to find the most efficient and effective solutions for complex systems. Finally, the integration of co-simulation with machine learning and artificial intelligence has provided new opportunities for the analysis and optimization of complex systems. Machine learning algorithms can be used to identify patterns and relationships in simulation data, allowing for the optimization of complex systems and the reduction of simulation time.

To sum up, co-simulation has made significant advancements in recent years, providing a more comprehensive representation of the behavior of complex systems, the integration of different simulation models and tools, cloud-based simulation platforms, integration with optimization techniques, and the integration of machine learning and artificial intelligence. These developments have made co-simulation an essential tool for the analysis and optimization of complex systems.

5.6 Reference Models of the Gunnerus Twin

Models and sub-models that represent the physical systems and behavior are the building blocks of a functional digital twin. In the co-simulation semantics, a model or sub-model is referred to as a component that can take the form of, for example, an FMU. These digital representations of real systems and behavior can be physics-based, data-based, algorithms, or only graphics for visualization purposes. The co-simulation approach for complex systems like a ship has its advantages in terms of modularity, modeling flexibility, and simulation efficiency. However, it always depends on the application to determine the complexity of the models to ensure both adequate simulation accuracy and efficiency. As mentioned above, the co-simulation framework allows for individual step sizes for each sub-simulator and orchestrates a network of multiple simulators using a different fixed or variable step size to enhance the computation speed. It also often comes with a price of comprised accuracy.

The maritime reference models are a collection of models and control systems representing typical maritime equipment. A set of example configurations are included

Table 5.1 Maritime reference models

Sub-system	Component	Tool/Code	User/OEM	Use case
Vessel	Hull Hull (+environment) PMAzimuth thruster Bow thruster Power plant Thruster drive	Simulink Vesim Vesim Vesim 20Sim 20Sim	NTNU Sintef Kongsberg Kongsberg NTNU NTNU	DP, Path-following, Docking, LARS, Palfinger Crane
Environment	Wind Current	Simulink	Kongsberg	DP, Path-following
Deck machinery	A-frame Crane Winch Winch (+controller)	20Sim 20Sim SimulationX Simulink	NTNU NTNU NTNU Kongsberg	Launch and Recovering (LARS), Palfinger crane
Controllers	DP controller Waypoint provider Passive observer Trajectory controller	JAVA	NTNU	Dynamic Positioning (DP), Path-following
Sensors	Wind sensor GNSS sensor VRU sensor Gyro sensor	C++	Kongsberg	Sensor communication configuration
Interfaces (NMEA)	Generic FMU Wind measurement (MWV) GNSS (GGA+GST) Gyro (HEDTH) VRU (SXN)	C++	Kongsberg	Sensor communication configuration

where models are connected and control systems run model-in-the-loop system simulations. The collection also includes examples of FMUs that provide network communication to send data to external applications, for instance, control systems. Table 5.1 includes a collection of components that represent the Gunnerus digital twin. They are developed by many different project partners using various methods and tools. Most of these components are open-sourced and support the FMI standard, except for the Sintef models a license is attached for authorised users. These components are used as reference models for the Gunnerus twinship in the use cases presented in the following chapters. As an example, the Gunnerus ship model and the Palfinger crane model for coupling in the ship-crane operation are described in the following subsections.

Furthermore, it is worth addressing that strong-coupled (a.k.a., tightly-coupled or tight coupling) systems are not naturally easy to be handled by co-simulation. As an example, the ship hull and the deck crane with a heavy payload whose dynamic inertial impacts to the ship cannot be neglected during operation. In this very specific case, the payload and the crane can be connected via a cable and treated as one component. The connection between the ship hull and the crane base can also be represented by a spring-damper system with extremely high stiffness and low damping coefficient. However, it requires extremely high computing resources for real-time simulation, which is critical for crane operations in this application. Alternatively, the acting forces from the crane, including the payload, can also be approximated as external forces applied to variant attacking points on the ship hull during operation. Using co-simulation is therefore a trade-off between the simulation accuracy and efficiency. This subsection describes the development of the Gunnerus ship model and the Palfiger crane model as examples of reference models for co-simulation used in the case studies [13, 112].

5.6.1 The Gunnerus Ship Model

The Gunnerus ship model in waves is a unified maneuvering and seakeeping model in six degrees of freedom (6-DoF) with DP control. The ship is modelded as a 6-DoF rigid body moving object where the fluid effect memories are modelled in a state space form. It combines the maneuvering and seakeeping ship models in a single code solved by Sintef's time-domain simulator VeSim [38]. VeSim is a 6-DOF rigid-body simulator, in which the manoeuvring problem is one component. The hydrodynamic derivatives are used to solve maneuvering problems, with outlines shown of the nature of each set of coefficients.

$$(M_{RB} + M_A)\dot{v} + C(v)v + D(v) + g(\eta) + \int_0^t h(t - \tau)v(t)\,dt = q \tag{5.16}$$

where M_{RB} is the rigid body mass, M_A is the added-mass, $C(v) = C_R B(v) + C_A(v)$ is a matrix function describing the generalised coriolis-centripetal forces, $D(v) = [X_d; Y_d; 0; 0; 0; N_d]^T$ is the vector of damping forces and moments, and $h(t)$ is a six by six matrix of impulse response functions, describing the force behaviour as calculated by SINTEF OCEAN's potential theory code, VERES.

VeSim has been developed as a combined manoeuvring and seakeeping simulator. Among many features, VeSim's capability to simulate the influence of ventilation effects on ship speed and powering makes it an outstanding tool. In VeSim, seakeeping is handled through the use of potential theory handled in the time domain. The manoeuvring model utilises the approaches from [167]. The manoeuvring equations in 3 degrees-of-freedom, which considers zero frequency added mass effect, added mass Coriolis-centripetal forces, linear lift and drag, cross flow drag. To improve the versatility of VeSim, SINTEF OCEAN has additional tools at hand which allow the manoeuvring model to be derived from advanced data sources, such as CFD simulations or PMM tests

For the Gunnerus digital twin ship, the ship model is wrapped and provided as a standalone co-simulation FMU, with the thruster models (two Azimuth thrusters and one tunnel thruster) as separate FMUs. External forces/moments are sent to the ship hull at any attacking points. These include the thrust forces, and acting forces by the crane and load as shown in Fig. 5.6. In order to couple the crane and the ship

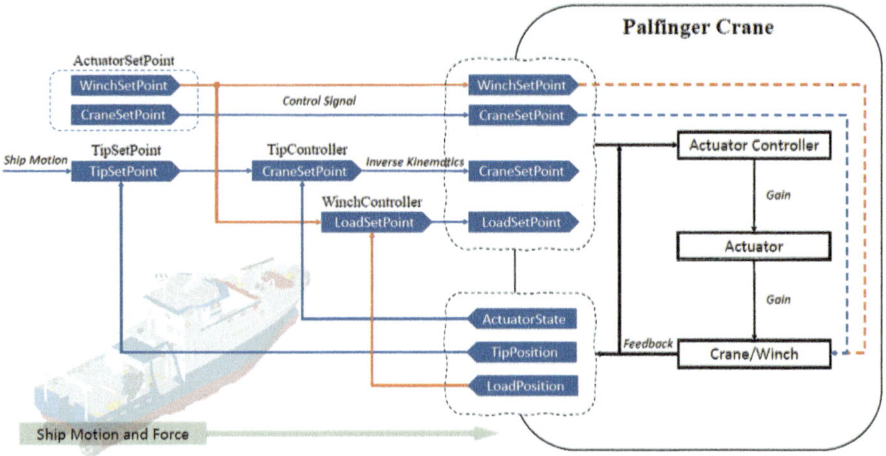

Fig. 5.6 Coupling of the Gunnerus ship model and the Palfinger crane model

using co-simulation, the crane position and the crane tip position where the forces from the load apply need to be calculated with respect to the ship motions.

5.6.2 *The Palfinger Crane Model*

The Palfinger crane on the Gunnerus vessel consists of three rotational joints and a translational joint. Following the Denavit-Hartenberg (D-H) convention, the crane's kinematic structure is described using five reference frames from the base to the tip, as shown in Fig. 5.7. It is to be noted that the last telescopic link l3 consists of eight segments. The blue lines represent the simplified line sketch for the links, and each link represents one rigid body of the crane. The green lines represent simplified geometrical relations for the transformation between the crane cylinders and the joints. The D-H table is presented as in Table 5.2.

We use the Lagrangian approach to derive the equations of motion for the multi-body crane. Lagrange's equations provide an elegant formulation of the dynamics of a mechanical system. It reduces the equations needed to describe the motion of the system using generalised coordinates instead of every single body with mass and inertia. It also provides a convenient form for modelling using the energy-based method, namely, the Bond Graph method [168]. A special type of bond graph called IC-field is used to represent the inertia, centrifugal and Coriolis terms of the model. We present the I and C matrix for the Palfinger crane below.

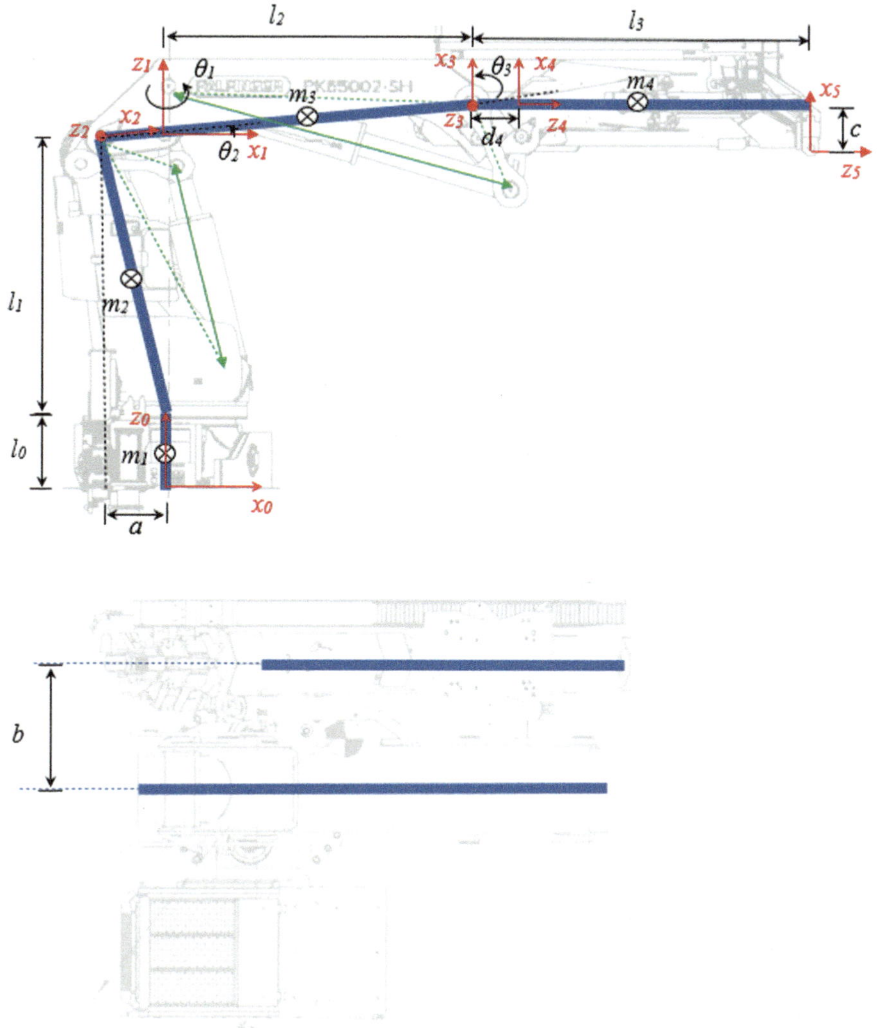

Fig. 5.7 Kinematics frames of the Palfinger crane by D-H method

$$I(\theta) = \begin{bmatrix} I_{11} & I_{12} & I_{13} & I_{14} \\ I_{21} & I_{22} & I_{23} & I_{24} \\ I_{31} & I_{32} & I_{33} & I_{34} \\ I_{41} & I_{42} & I_{43} & I_{44} \end{bmatrix}, C(\theta, \dot{\theta}) = \sum_{k=1}^{n} \Gamma_{ijk}\dot{\theta} = \begin{bmatrix} C_{11} & C_{12} & C_{13} & C_{14} \\ C_{21} & C_{22} & C_{23} & C_{24} \\ C_{31} & C_{32} & C_{33} & C_{34} \\ C_{41} & C_{42} & C_{43} & C_{44} \end{bmatrix}$$

The non-zero elements are given as follows, where s and c denote sin and cos.

Table 5.2 D-H table for the Palfinger crane

i	a_{i-1}	α_{i-1}	d_i	θ_i
1	0	0	l_0+l_1	θ_1
2	a(0.38m)	90	0	θ_2
3	l_2	0	b(−0.6m)	θ_3
4	0	90	d_4	0
5	c(0.5m)	0	l_3	0

$$
\begin{aligned}
I_{11} &= I_{y2}s_2^2 + I_{y3}s_{23}^2 + I_{y4}s_{23}^2 + I_{11} + I_{z2}c_2^2 + I_{z3}c_{23}^2 + I_{z4}c_{23}^2 \\
&\quad + m_2r_1^2c_2^2 + m_3(l_1c_2 + r_2c_{23})^2 + m_4(l_1c_2 + (r_3 + d_4)c_{23})^2 \\
I_{12} &= I_{x2} + I_{x3} + I_{x4} + m_3l_1^2 + m_4l_1^2 + m_2r_1^2 + m_3r_2^2 \\
&\quad + m_4(r_3 + d_4)^2 + 2m_3l_1r_2c_3 + 2m_4l_1(r_3 + d_4)c_3 \\
I_{23} &= I_{x3} + I_{x4} + m_3r_2^2 + m_4(r_3 + d_4)^2 \\
&\quad + m_3l_1r_2c_3 + m_4l_1(r_3 + d_4)c_3 \\
I_{24} &= I_{x4} + m_4(r_3 + d_4)^2 + m_4l_1(r_3 + d_4)c3 \\
I_{32} &= I_{23} \\
I_{33} &= I_{x3} + I_{x4} + m_3r_2^2 + m_4(r_3 + d_4)^2 \\
I_{34} &= I_{x4} + m_4(r_3 + d_4)^2 \\
I_{42} &= I_{24} \\
I_{44} &= I_{43} = I_{34} \\
\Gamma_{112} &= (I_{y2} - I_{z2} - m_2r_1^2)c_2s_2 + (I_{y3} - I_{z3} + I_{y4} - I_{z4})c_{23}s_{23} \\
&\quad - m_3(l_1c_2 + r_2c_{23})(l_1s_2 + r_2s_{23}) \\
&\quad - m_4(l_1c_2 + (r_3 + d_4)c_{23})(l_1s_2 + (r_3 + d_4)s_{23}) \\
\Gamma_{113} &= (I_{y3} - I_{z3} + I_{y4} - I_{z4})c_{23}s_{23} - m_3r_2s_{23}(l_1c_2 + r_2c_{23}) \\
&\quad - m_4(r_3 + d_4)s_{23}(l_1c_2 + (r_3 + d_4)c_{23}) \\
\Gamma_{121} &= \Gamma_{112} \\
\Gamma_{131} &= \Gamma_{113} \\
\Gamma_{211} &= -\Gamma_{121} = -\Gamma_{112} \\
\Gamma_{223} &= -m_3l_1r_2s_3 - m_4l_1(r_3 + d_4)s_3 \\
\Gamma_{232} &= \Gamma_{223} \\
\Gamma_{233} &= \Gamma_{232} = \Gamma_{223} \\
\Gamma_{311} &= -\Gamma_{131} = -\Gamma_{113} \\
\Gamma_{322} &= -\Gamma_{233} = -\Gamma_{232} = -\Gamma_{223}
\end{aligned}
$$

$$\Gamma_{141} = m_4(l_1c_2 + (r_3 + d_4)c_{23})c_{23}$$
$$\Gamma_{242} = m_4(4_3 + d_4) + m_4l_1d_4c_3$$
$$\Gamma_{243} = m_4(r_3 + d_4) + \frac{1}{2}m_4l_1d_4c_3 - \frac{1}{2}l_1(r_3 + d_4)c_3s_3$$
$$\Gamma_{244} = 2\Gamma_{242}$$
$$\Gamma_{342} = m_4(r_3 + d_4) + \frac{1}{2}m_4l_1d_4c_3 + \frac{1}{2}l_1(r_3 + d_4)c_3s_3$$
$$\Gamma_{343} = m_3(r_3 + d_4)$$
$$\Gamma_{344} = 2\Gamma_{343}$$
$$\Gamma_{444} = \Gamma_{343}$$
$$\Gamma_{411} = -\Gamma_{141}$$
$$\Gamma_{422} = -\Gamma_{242}$$
$$\Gamma_{423} = -\Gamma_{444} = -\Gamma_{343}$$
$$\Gamma_{424} = -\Gamma_{423}$$
$$\Gamma_{432} = -\Gamma_{342}$$
$$\Gamma_{433} = \Gamma_{423} = -\Gamma_{444} = -\Gamma_{343}$$

5.7 Summary

In this chapter, we discussed the applications of physics-based and data-driven modeling for MDOS, especially, hybrid modeling as and co-simulation as the fundamentals for digital twin development. These techniques are essential components of MDOS development and can be used to create and simulate a more accurate representation of the system. Specifically, the development of digital twin systems has made significant advancements in recent years, providing a virtual representation of the systems, with integration of data-driven and model-driven approaches, cloud-based simulation platforms, integration with optimization techniques, and integration of machine learning and artificial intelligence. These developments have enabled digital twin systems as essential tools for condition monitoring, analysis, optimization, prognosis and predictions.

Digital twin systems are virtual replicas of physical systems that allow for real-time monitoring, analysis, and optimization of their operation. They are used in a wide range of industries, including manufacturing, aerospace, and energy. OOM as a modeling technique offers a convenient and useful approach for domain-specific modeling, as it allows engineers to create accurate and detailed representations of the physical system being modeled and for the interfacing of different simulation models. It is up to the user to determine the complexity of the domain models and the level of co-simulation models for coupling, with considerations in simulation purpose, operation scenario, intellectual property rights, etc.

Following these above mentioned guidelines, modeling for the Gunnerus ship and the Palfinger crane was described. The coupling of the components models using co-simulation was presented. Case studies using the Gunnerus ship models are presented in the following chapters.

Yingguang Chu is an Associate Professor with University of Southern Denmark (SDU). Dr. Chu received his Ph.D. degree from Norwegian University of Science and Technology (NTNU) in 2018. From 2018 to 2022, he worked as a research scientist with Sintef, Norway, on digital twin system development. His current research fields and interests at SDU include mechanical engineering, hydraulic power systems, and modeling and simulation.

Guoyuan Li received the Ph.D. degree in computer science from the Department of Informatics, Institute of Technical Aspects of Multimodal Systems, University of Hamburg, Hamburg, Germany, in 2013. In 2014, he joined the Department of Ocean Operations and Civil Engineering, Intelligent Systems Laboratory, Norwegian University of Science and Technology (NTNU), Ålesund, Norway, where he is currently a Professor in ship intelligence. He has published more than 150 articles in the areas of his research interests, including the modeling and simulation of ship motion, autonomous navigation, intelligent control, optimization algorithms, and the locomotion control of bioinspired robots.

Vilmar Æsøy received the Ph.D. degree from the Norwegian University of Science and Technology (NTNU) for his research on natural gas ignition and combustion through experimental investigations and numerical simulations in 1996. Since 2002, he has been employed in teaching with the Ålesund University College, where he is also developing and teaching courses in marine product and systems design on bachelor's and master's level. In 2010, he received the green ship machinery professorship at NTNU. His current research interest includes energy and environmental technology, with focus on combustion engines and the need for more environmental friendly and energy efficient systems.

Houxiang Zhang received the Ph.D. degree in mechanical and electronic engineering from Beihang University in 2003 and the Habilitation in informatics from the University of Hamburg, Germany in 2011. He is currently a full Professor with the Department of Ocean Operations and Civil Engineering, Faculty of Engineering, Norwegian University of Science and Technology (NTNU), Norway. From 2011 to 2016, he also hold a Norwegian National GIFT Professorship on product and system design funded by the Norwegian Maritime Centre of Expertise. He has engaged into two main research areas, including control, optimization and AI application especially on autonomous vehicle; and marine automation, digitalization and ship intelligence. He has applied for and coordinated more than 30 projects supported by Norwegian Research Council (NFR), German Research Council (DFG), EU, and industry. In these areas, he has published over 300 journals and conference papers as the author or coauthor. He was nominated and selected as a member of the Norwegian Academy of Technological Sciences in 2019, and a member of Academy of the Royal Norwegian Society of Sciences and Letters in 2024.

Chapter 6
Enabling Tools for Maritime Operation Support

Chunlin Wang, Baiheng Wu, Guoyuan Li, Hans Petter Hildre, and Houxiang Zhang

The digital twin of a surface ship is the virtual representation of its physical counterpart with bi-directional communication. To achieve operation support based on digital twins, such as early warning, prediction and path optimization, enabling technologies are needed. Enabling tools in such a context could serve as middle-ware and play an important role in combination with raw data and human factors to support advanced marine operations. Figure 6.1 illustrates a scheme of enabling tools for maritime operation support in our digital twin system. It roughly contains two parts, i.e., data processing and analysis, and human-in-the-loop development. The former focuses on dealing with data flow or data segments from the database for deep insights into data associations, clusters, and dependencies, thus simplifying and optimizing data-driven models; the latter puts efforts on human-machine interaction, with a focus on human-related data sampling and collection, operational behavior analysis, to human-centered decision support. The following sections describe relevant enabling technologies/tools for building decision support for marine operations.

6.1 Data Processing and Analysis Capability

Maritime operations are inevitably influenced by the winds, waves, sea currents or other perturbations at sea. Predicting these operations based on historical and real-time data on ship status is thus of great concern in terms of ship safety. However, it is challenging to construct predictive model, especially since huge amount of data the raw sensor data and operational states may contain noisy, discontinuous and redundant information. Therefore, it is necessary to process and purify the data so that their affection on further analysis and modelling can be minimized.

The detailed modelling workflow is shown in Fig. 6.2. The first step is data collection. After that, data preprocessing is carried out to purify the marine operation data for the follow-up model construction. Prior to modelling, the model structure

H. Zhang et al., *Digital Twins for Vessel Life Cycle Service*,
https://doi.org/10.1007/978-981-97-4768-9_6

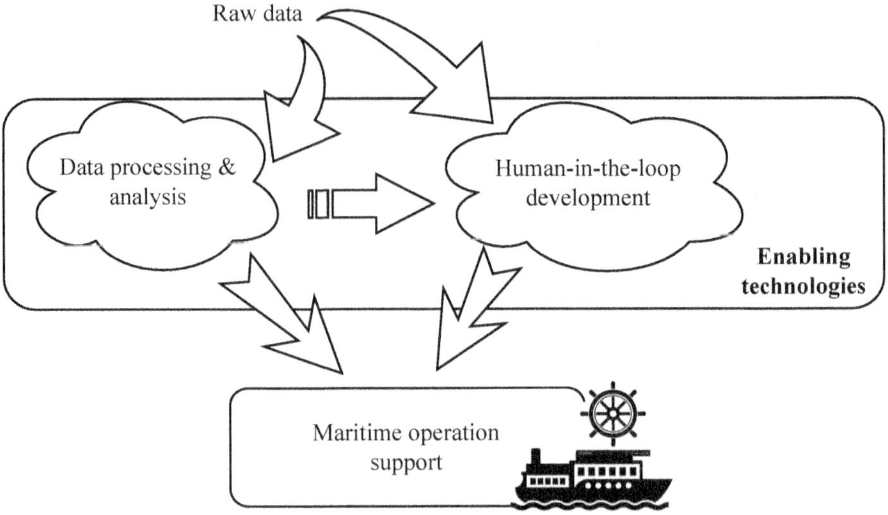

Fig. 6.1 Enabling technologies for digital twin-based onboard support of marine operation

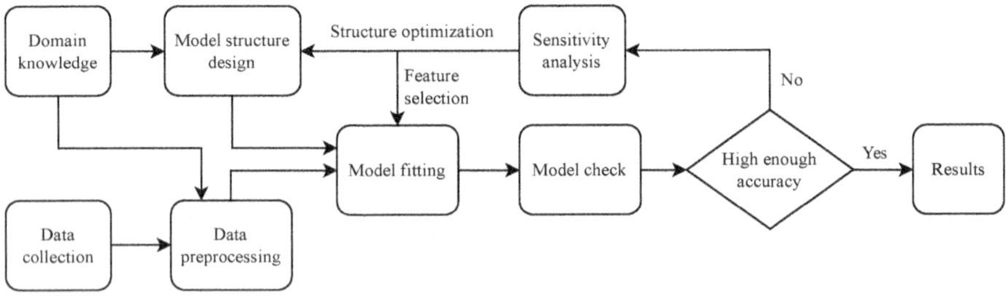

Fig. 6.2 The flow chart of marine operation data modelling

is designed according to the data characteristics and the research objective. Therein, domain knowledge plays a key role in both data preprocessing and model structure design. Next, the processed data is used to train model parameters to obtain an appropriate model. In the process of model check, if the model performance cannot achieve an expected accuracy, sensitivity analysis (SA) is used to evaluate the model structure and rank relevant parameters. It could be helpful to select important input features to improve model performance. As a result, the models characterized by optimized structure and input features can be constructed and saved for use in high-level applications.

In the modelling workflow, data quality and model structure play key roles in model accuracy. In order to fit the model to the data, the model is designed more and more complex, resulting in a potential increase of overall uncertainty to the output, known as uncertainty cascade [169]. SA is considered a powerful enabling technology that is beneficial to model structure optimization. The following section will introduce the usage of data preprocessing and SA.

6.1.1 Data Preprocessing

In recent decades, the scale of collection and storage of marine operation data has increased significantly, forming big data in marine operations. Big data contains abundant and useful knowledge for decision support for marine operations. However, due to the significant characteristics of big data such as high dimensionality, large data volume, and noise pollution, it is extremely challenging to directly discover potential knowledge from big data [170]. Therefore, data preprocessing is an important foundation for subsequent data analysis and knowledge discovery. The main tasks of data preprocessing contain data cleaning, data reduction, data transformation, and data augment.

1. ***Data cleaning***: Data cleaning aims to enhance the quality of data. Considering the fact that marine operation data is sometimes partially missing or simply wrong due to human error and environmental disturbances, one can make use of the following technologies to identify incomplete, inaccurate, duplicated, irrelevant, or null values in the dataset:

   (a) *Missing data imputation*: The missing data problems can be addressed using univariate and multivariate methods. Univariate methods infer the missing values by means or medians of variables, forward/backward method, and moving average method. Univariate imputation methods may not capture cross-sectional and temporal correlations in the data and thus are only suitable for the cases with low missing data ratio [171]. Multivariate methods can estimate the missing data values accurately, including k-nearest neighbor (KNN) and regression/interpolation-based methods. KNN can find the means of k most similar data samples to replace the missing data [172]. It can achieve a better performance when the missing data ratio grows up to 15% [173]. Regression/interpolation-based methods can fit a regression model with respect to time and estimate the missing data at the given time moment [174]. Commonly used methods are linear interpolation, polynomial interpolation, and ML methods such as support vector machine (SVM) and artificial neural network (ANN).

   (b) *Outlier detection*: Outlier detection removes the abnormal values by statistical and clustering-based methods. The quantile is a classic statistical method to filter abnormal data. If the data fall beyond the range of $[Q_1 - 1.5(Q_3 - Q_1), Q_3 + 1.5(Q_3 - Q_1)]$, where Q_1 and Q_3 are the first and third quantile, it is identified as outliers [175]. Another popular method is clustering such as KNN, Density-Based Spatial Clustering of Applications with Noise (DBSCAN), and Isolation forest. They identify outliers if the data are located outside of clusters. Compared with statistical methods, the clustering method can consider the similarity of data samples in the high-dimensional feature space.

2. ***Data reduction***: Marine operation data contains a large amount of redundant information. Therefore, data reduction should be applied to extract important data information. It could be carried out in two directions, that is, row-wise for data sample reduction and column-wise for data variable reduction [171].
   For row-wise data reduction, sampling technologies are typically used to reselect as much data sample as the practitioner wants. For example, random sampling selects data samples for each input variable independently and uniformly from the entire range. It does not guarantee that each interval can be covered in the sample. To address this problem, hypercube sampling was proposed to divide the range of each input variable into equal intervals and select values from each interval [176]. Another method is the stratified sampling which can maintain the proportion of data samples corresponding to different categories.
   For column-wise data reduction, statistical feature selection methods can be used to remove redundant data variables. The commonly used approaches are divided into three categories: the filter approach, the wrapper approach, and the embedded approach [177]. Feature extraction methods are also used for variable reduction, such as principle component analysis, deep autoencoder, and SA [171].
3. ***Data scaling***: Marine operation data variables have different scales, which could degrade the performance of the constructed models. To address this problem, data scaling is used to normalize the variable values to a rational range. The widely used methods are max-min normalisation and z-score standardization. Max-min normalization algorithm can transform the values to the range of [0, 1]. However, it is sensitive to outliers that have a large scale as they can change the variable range. By contrast, the z-score standardisation method is less influenced by outliers and can handle units with different scales [178].
4. ***Data augmentation***: Marine operation data may contain an unequal distribution of classes, where the number of samples in the majority class is far more than that in the minority class [179]. The imbalanced training data can affect the performance of the constructed data-driven model. Therefore, data augmentation technologies, such as oversampling and undersampling, are often used to obtain more balanced distributions [180]. The undersampling method can remove some samples from the majority class. However, this method could lead to an under-fitting problem. The oversampling method duplicates data samples in the minority class until both classes are balanced. The common approaches are random oversampling, and cluster-based oversampling [181].

6.1.2 *Sensitivity Analysis*

Sensitivity analysis (SA) is commonly defined as "*The study of how uncertainty in the output of a model (numerical or otherwise) can be apportioned to different sources of uncertainty in the model input*" [182]. SA is accomplished by answering the question of which model inputs influence the model output of interest. Models can take various forms such as systems, physical processes, functions, mathematical models, and data-driven models. Model inputs commonly refer to factors that involve model parameters, data features, forcing variables, boundary and initial conditions, model structural configurations, assumptions, and constraints [183]. Model outputs represent a summary variable that is obtained by objective functions (such as a production or cost function in cost-benefit analysis) and prediction functions (such as the value of a target variable or its average over a spatial and temporal domain) [184, 185]. Given its desirable capability of factor importance analysis, recent decades have seen ever-increasing interest to unleash SA capabilities to address the fundamental overarching problems of systems analysis and modelling in engineering [186], technology [187], social sciences [188], earth science [189], etc.

How can SA benefit scientific investigation and decision-making? The possible applications are shown as follows: (1) exploring causalities and how different choices, hypotheses, parameters, conditions, and their combinations affect the system response of interest [190], which can guide practitioners toward making the right decision to implement the expected outcomes further; (2) conducting dimensionality reduction to simplify and optimize the model by fixing or removing non-influencing factors [191]; (3) making data worth assessment to reduce the targeted uncertainty the most [192, 193]; (4) providing decision support to quantify the sensitivity of different decision options, constraints, and assumptions to an expected outcome [194]. More cross-disciplinary applications of SA can be referred to [195].

Overview of SA Approaches

Different SA approaches have their own advantages and limitations. Consequently, the use of SA depends on the modelling domain and the specific application. Therefore, it is necessary to revisit the fundamentals of SA methods and their pros & cons and probe the ways toward marine digitalization. In the following, we will introduce the SA-related approaches as shown in Fig. 6.3. All SA approaches are mainly categorized into two groups: local SA (LSA) and global SA (GSA). LSA is performed in a local input space around a point of interest while GSA considers the entire space variability of the input factors. The common methods of LSA are graph-based SA and perturbation & derivative-based SA. Most proposed methods are GSA including multiple-starts perturbation-based SA, regional SA (RSA), correlation & regression-based SA, meta model-based SA, variance-based SA, distribution-based SA, and variogram-based SA. These methods are mainly used for the three settings (or purposes) of SA including factor ranking, factor fixing, and factor mapping [196].

Factor ranking (or factor prioritization) aims to rank the input factors based on their sensitivity indices. Factor fixing (or factor screening) is to identify uninfluential factors and fix them at constant values. Factor mapping can determine the region of input variation space that affects the output values significantly.

In general, sensitivity indices cannot be computed analytically as a result of the complex relationship between model inputs and outputs and thus they are numerically approximated via simulations or model evaluations with the resampled data [184]. The main sampling strategies are one-at-a-time (OAT) and all-at-a-time (AAT). In the OAT method, an input factor is varied over the sampling space while other factors remain fixed. For AAT, all input factors are varied simultaneously, and hence the sensitivity of each factor contains the direct influence of this factor and the joint influence with other factors. After sampling, the obtained sample is fed into the model to make model evaluations. Note that 'M' denotes the number of data features in Fig. 6.3.

- **Graph-based SA**: It is an LSA method, which can represent the sensitivity of factors intuitively. The common graphical representation methods are Tornado diagrams, scatter plots, and spider plots. Tornado diagrams quantitatively represent the sensitivity indices as horizontal bars, sorted from the largest to the smallest [197]. Spider plots can show the sensitivity of inputs and reveal whether the relationship between inputs and outputs is linear or non-linear [198]. Likewise, scatter plots can observe the relationship between model inputs and outputs [199]. However, scatter plots are a qualitative SA while spider plots represent the sensitivity measures quantitatively and qualitatively.
- **Perturbation & derivative-based SA**: The simplest way of SA is to compare two sets of output values that are obtained by model evaluations of nominal inputs and the perturbation of inputs [200]. The sensitivity of each input factor (x_i) at $\bar{x}$ is measured with $S_i(\bar{x}) = \frac{\partial y}{\partial x_i}|_{\bar{x}}$. It is usually approximated by an analytic form $\hat{S}_i(\bar{x}) = \frac{f(\bar{x}_1,\ldots,\bar{x}_i+\Delta_i,\ldots,\bar{x}_M)-f(\bar{x}_1,\ldots,\bar{x}_i,\ldots,\bar{x}_M)}{\Delta_i}$, where $y = f(x)$. This method only requires $M+1$ model evaluations to compute the sensitivity of M factors. Therefore, derivative-based SA has a low computational cost. However, it is limited to conducting LSA without considering the variation of the entire input space.
- **Multiple-starts perturbation-based SA**: This method is an extension of the perturbation method from multiple points $\bar{x}^j$ within the feasible input space. The global sensitivity is estimated by aggregating these individual sensitivities [184]. The elementary effect test (EET), also called Morris [201], is the widely used multiple-starts perturbation-based SA method [196]. It is taken as a GSA method because the baseline changes in every step and the final sensitivity measures are calculated by averaging at different points of the input space (or levels) [202]. Given l levels and n model evaluations, there are l^n points in the grid, from which a subsample of r points is drawn at random [203]. The elementary effect (EE) (or finite difference) is the averaged value of r points computed with

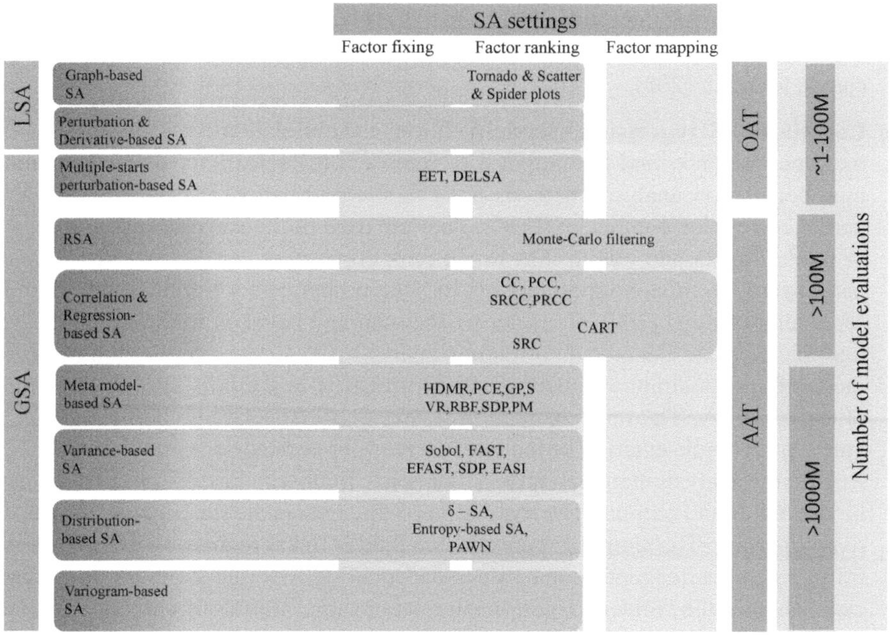

Fig. 6.3 Taxonomy of sensitivity analysis approaches

$$S_i = \frac{1}{r}\sum_{j=1}^{r} \frac{f(\bar{x}_1^j, \ldots, \bar{x}_i^j + \Delta_i^j, \ldots, \bar{x}_M^j) - f(\bar{x}_1^j, \ldots, \bar{x}_i^j, \ldots, \bar{x}_M^j)}{\Delta_i^j} \quad (6.1)$$

Since the computation of EE of M input factors requires $r(M+1)$ model evaluations. Therefore, it is suitable for situations where the computational cost of running the model is high or there are many input factors to the model [184].
The distributed evaluation of LSA (DELSA) is a derivative-based local method, in which multiple evaluations of local parameter sensitivity are distributed across parameter space [204]. It can identify the important parameters and estimate how model parameter importance changes as parameter values change. As the local sensitivity indices are computed throughout the parameter space, it is taken as a GSA method. A more detailed description can be referred to [204].

- **RSA**: Also called Monte Carlo filtering, was first proposed in [205, 206]. It aims at determining regions of input samples corresponding to the particular model output. The input samples are split into two groups based on whether the associated model output satisfies a predefined condition, e.g., above/below a threshold. For the purpose of factor mapping, the two input sets are compared with empirical cumulative distribution function (CDF), for instance, by highlighting the reduced range of variability of the inputs within the group [207]. RSA is suitable for any type of model output (numerical or non-numerical), as long as the split criterion can be given. However, the choice of split criterion is subjective for the practitioners. In other words, there is no standard to choose the threshold. More detailed discussion can be found in [208].

- **Correlation & regression based SA**: It uses the correlation coefficients between the input factor x_i and the output y to represent the sensitivity of x_i. The common correlation analysis methods are Pearson correlation coefficient (CC) and partial correlation coefficient (PCC). They are used for linear relationship analysis of model inputs and output. Spearman rank correlation coefficient (SRCC) and partial rank correlation coefficient (PRCC) are applied for nonlinear but monotonic relationships [209]. Considering the mapping between multiple inputs and multiple outputs, Canonical correlation analysis (CCA) was proposed to estimate the coefficients simultaneously [210]. Correlation analysis is similar to scatter plots-based SA. The only distinction is that those correlation coefficients are a quantitative representation of the sensitivity of input factors.
Regression analysis methods can fit a function to represent the relationship between model inputs and outputs. The regression coefficients in the function are the sensitivity measure. The simplest regression method is linear regression which aims at fitting a straight line represented by a linear equation. The linear regression method cannot handle the problem of non-linear relationships. Therefore, classification and regression trees (CART) are proposed to analyze the nonlinear regression problems [211, 212]. It is also suitable for non-numerical inputs and outputs and can be used for both factor ranking and mapping.
- **Variance-based SA**: Correlation & regression-based SA is not sufficient to analyze parameters' sensitivity in complex models such as the interactions of parameters. To address this problem, variance decomposition was introduced to analyze the total effect of parameters [213]. The variance decomposition based method is to compute the sensitivity of input factors by decomposing the variance of outputs to each factor [182]. According to [184], The variance-based SA is conducted based on the three basic assumptions: (1) inputs are stochastic variables that follow a probability distribution; (2) the total variance of the model output can represent the output uncertainty; (3) the sensitivity index of an input factor can be regarded as its contribution of the output variance.
The variance-based SA can compute the first-order and total-order indices of input factors. First-order indices refer to the main contribution to the output variations from individual input factors. It is computed by $S_i = \frac{V_i}{V(Y)}$, in which S_i is the sensitivity index of ith factor; V_i is the variance of input factor x_i; $V(Y)$ is the total variance of output Y. Total-order indices calculate the total effects including the main effect and the interactions among input factors. They are obtained by $S_{Ti} = \frac{V_i + \cdots + V_{ij} + \cdots + V_{1,\ldots,M}}{V(Y)} = 1 - \frac{V_{\sim i}}{V(Y)}$, where $V_{\sim i}$ represents the total variance contribution of remaining parameters to Y given x_i. The most representative method is Sobol's method which was proposed in [213]. Sobol's method decomposes the variance via Monte Carlo methods, which requires a large number of model simulations. Another classic variance-based method is the Fourier Amplitude Sensitivity

Test (FAST) method. The FAST method uses sinusoidal functions of different frequencies to explore the input space and then applies spectral analysis to apportion the variance [214]. The improved version of FAST is called eFAST which can reduce the computational cost of SA [215].

- **Meta model based SA**: Also called surrogate model-assisted SA, uses a surrogate model to approximate the complex simulation models in the scientific and engineering disciplines. It is a two-stage method: (1) creating a meta model with non-parametric regression methods; (2) computing sensitivity indices using the variance-based method. The commonly used meta models contain the polynomial regression model (PRM), high-dimensional model representation (HDMR), state-dependent parameter (SDP), polynomial chaos expansion (PCE), Kriging/Gaussian Process (GP), and Support Vector Regression (SVR).
  The PRM, also known as the response surface method, is the simplest surrogate model for SA [216]. It estimates the sensitivity indices by using the regression coefficients directly. Although it is easy to implement, it is limited to building models with high dimensionality as the training sample size must be greater than the number of unknown coefficients, i.e. $N > (M + 1)(M - 1)$. In addition, the modelling accuracy can not be guaranteed for highly nonlinear and multimodal problems [217]. The HDMR can quantify the independent and cooperative contributions of the inputs on the model output [218]. It is capable of obtaining the sensitivity indices of high-dimensional inputs. Two common HDMR models are Cut-HDMR and AVOVA-HDMR [219, 220].
  The SDP method is a signal processing and analysis tool. It was firstly applied as a surrogate model to obtain the variance-based sensitivity indices by estimating the SDP coefficients [221]. The Kriging model, also called GP, is to find the best linear unbiased function that is composed of the polynomial basis function and a Gaussian process. The ordinary Kriging model can be transformed into a sum of multivariate tensor product basis functions and the Sobol indices can be obtained analytically [216, 222]. The SVR is designed to fit the regression model and its shape is consistent with the PCE model. Therefore, variance-based sensitivity indices can be calculated by post-processing the SVR coefficients [223]. The SVR has a better generalization performance than other meta-model methods, but its performance is sensitive to the hyper-parameters [217]. Similar to SVR, a radial basis function (RBF) model can also be represented by a sum of multivariate tensor product basis functions that can be used to compute sensitivity indices analytically [224].
  The above meta models can obtain sensitivity indices by computing the regression coefficients analytically. Another way is to sample input samples from input space and then obtain output by the evaluation of the constructed meta-models. After the collection of model inputs and outputs, variance-based or distribution-based SA methods can be used to estimate the sensitivity indices of all input factors.

- **Distribution-based SA**: The variance-based SA is insufficient to estimate sensitivity indices when model outputs are highly skewed and multi-modal [225]. Therefore, distribution-based SA was proposed to consider the changes in the whole probability distribution, not just the variance. For example, the Borgonovo δ estimates sensitivity indices by computing the expected shift in the probability distribution of the model output at a given input factor. This method is computationally expensive because of the unconditional and conditional probability distribution function (PDF) estimation and a double integral computation [226]. In order to improve the efficiency of PDF estimation, the maximum entropy technology is applied to obtain unconditional and conditional PDFs without extra model evaluations [227]. However, PDF estimation still requires a high computational cost and makes it not applicable to high-dimensionality problems. Francesca et al. came up with a novel SA method called 'PAWN' that characterizes the output distribution by its CDF instead of PDF [228]. One advantage of PAWN is that it hugely reduces computational cost because there is no need to compute unknown parameters for the approximation of empirical CDF. Another advantage is that sensitivity indices can be easily obtained, by considering either the entire range of variation of the model output or a sub-range.

- **Variogram-based SA**: The variogram-based SA (VARS) is a new technique used to obtain the sensitivity of model inputs by analyzing the behavior of the variogram. In practice, model output is not always randomly distributed [229]. VARS can consider the spatially ordered (covariance) structure in the input structure, while conventional SA methods may not completely capture the spatial and correlated relationship [183]. In addition, VARS has the advantage of high computational efficiency and an intuitive and comprehensive characterization of sensitivity across the full spectrum of scales in the factor space. However, VARS cannot be used in any given data set. To address this problem, a modified VARS is proposed to approximate the anisotropic variogram structure of the underlying (but unknown) response surface, when only a (small) sample of the input[U+2010] output space is available [230].

Applications of SA for Marine Operation Digitalization

Marine operation digitalization refers to the creation of digital models to make behavior predictions, early warning, maintenance, and optimization for physical assets and systems. To simulate the true behavior of systems, these digital models are constructed with complicated internal mechanics and assumptions. For example, a lot of parameters are introduced into the models to reduce the model incompleteness and thus represent the system behavior adequately. Meanwhile, the uncertainty in these parameters could be propagated to model outputs, which can result in a large model error [231]. The potential of SA in promoting the development of marine operation digitalization will be elaborated from different perspectives below.

- **Model reduction**: digital models, such as mathematical models, are becoming more and more complex with the increasing number of model parameters. Some of these parameters have a small influence on the model output and can be removed to simplify the constructed model. For example, SA was applied to analyze the importance of 19 coefficients in the maneuvering mathematical model [232, 233]. The influential coefficients are retained and used to build a reduced model that achieves the same performance as the original model.

- **Feature selection**: to advance the development of digital models, ML and DL are used to train digital models using marine operation data. The notable characteristic of marine operation data is high dimensionality which causes the curse of dimensionality to the ML model training. To reduce the computational cost of model training, some unimportant input features should be picked and removed from the training dataset. To this end, SA is a useful enabling tool to rank data features. For example, it was employed to identify the important input factors in the data-driven model of ship motion and use the picked important input features to construct a compact ship motion model [234].
- **Model uncertainty quantification**: models suffer from uncertainty caused by uncertain factors. To quantify model output uncertainty, uncertain factors, such as model parameters or coefficients, and input features, need resampling with a high frequency, which exacerbates the difficulty of model evaluations. Hence, it is necessary to identify the key drivers of model uncertainty, in order to reduce the burden of uncertainty quantification. For example, at first, SA was applied to find out the important parameters of the modal parameters identification model. After that, resampling was carried out based on the picked parameters to implement rapid uncertainty quantification [235].
- **System optimization**: the system design problem is formulated in the form of physical design parameters, objective functions, and constraints. Typically, objective functions are functions of the defined parameters and need to be optimized to obtain the optimal system responses [236]. However, a large number of parameters increase the difficulty of system optimization. Therefore, feature selection becomes necessary to pick up the important parameters that have the dominant control over the system response. For example, the ship hull form optimization involves dozens of parameters to be optimized to improve performance parameters such as resistance, seakeeping, and stability [237]. To handle the curse of dimensionality, variance-based SA was applied to remove uninfluential variables to improve optimization efficiency [238].
- **Risk assessment**: Marine operations suffer from the risk of uncertainties caused by environments, human decisions, etc. A better understanding of which factors have a dominant influence on operational tasks is critical to enable a safer marine operation. For example, busy marine traffic brings many risk factors and threatens the safety of ship navigation in restricted waterways. Therefore, understanding the interaction of uncertain factors with ship navigation risk is important for decision support and collision avoidance. To this end, SA was applied to investigate which factors cause an impact on the navigation risk of Liquefied Natural Gas carriers. Such a study is valuable for the guidance of improving the efficiency and safety strategies implemented in restricted waterways [239].

6.2 Human-in-the-Loop (HITL) Studies

Besides the enabling tools used for data processing and analysis, HITL as an enabling technology was also considered for digital twin development. In recent years, there has been significant advancement in the development of Maritime Autonomous Surface Ships (MASS) in both academic and industrial domains, including technical solutions and legal regulations. The Maritime Safety Committee (MSC) of the International Maritime Organization (IMO) [240] organizes ship autonomy in four degrees:

(i) Crewed ship with automated processes and decision support;
(ii) Remotely controlled ship with seafarers on board;
(iii) Remotely controlled ship without seafarers on board, and
(iv) Fully autonomous ship.

According to IMO, human navigators will still play varying degrees of roles in ship navigation, and conventional navigation will continue to be predominant in most implementations, including MASS in categories (i)–(iii). The IMO plans to enhance the development and application of MASS between 2024 and 2028, with the application of MASS estimated to occur gradually between 2030 and 2060, signifying a period of coexistence and collaboration between human navigators and machine intelligence.

Figure 6.4 illustrates the framework of synthesis research of HITL navigational operations. Based on a variety of experimental platforms, the HITL research can be categorized in two ways: human-centered intelligent ship bridge system, which focuses on the navigators' physical and psychological activities while on assignment; and onboard expertise data study, which leverages maneuvering and operational data to learn and summarize occupational expertise and experience. Both ways lead to

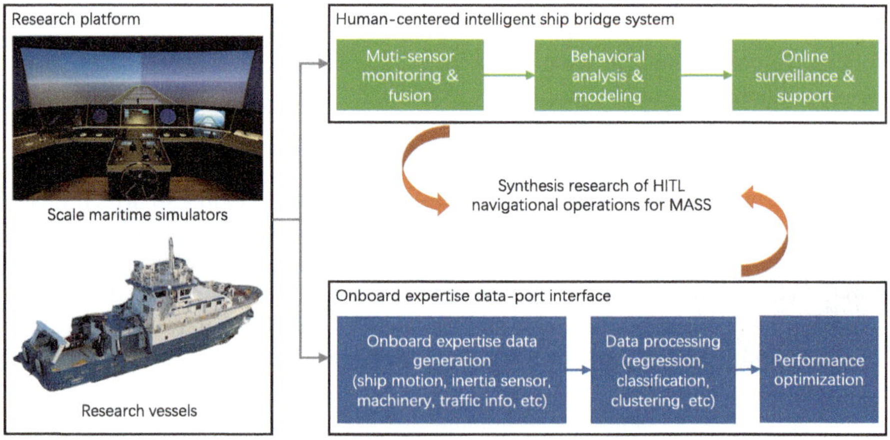

Fig. 6.4 Illustration of synthesis research of HITL navigational operations towards MASS

similar goals: promoting onboard navigational performance and providing decision support tools. All components in the loop illustrated in Fig. 6.4 contextualize the idea of synthesis study of HITL in maritime navigation [12].

6.2.1 Human Behavioral Data from Ship Bridges

Figure 6.5 depicts the workflow of an intelligent ship-bridge system in general. In this framework, there are the following parts:

- Executives: The human navigators are also the bearers to be monitored and recorded by the multi-sensor system;
- Multi-sensor monitoring: More channels of sensors are expected in the monitoring systems to provide cross-explain to the navigators' behaviors; However, the more sensors equipped, the more influence might be caused on the navigators' performance. In this context, non-intrusive solutions are preferable. If not available, the number of sensor channels should be controlled;
- Behavior modeling: Data interpretation and data mining by machine learning algorithms to explore the potential navigational mechanisms;
- Derived functions: Some passive functions without intervention to the control loop are developed based on the modeled navigational behaviors.

Tools for human-related data collection can be sorted as intrusive and non-intrusive. To record electroencephalogram (EEG) data, a wearable brain data hardware is inevitable [241]. When it comes to visual attention tracking, two types of solutions are available: the first type is intrusive, e.g., wearable eye trackers [242, 243], which is popular and versatile in conducting various research items; the other type is non-intrusive camera-based (RGB/RGBD) algorithms, which have been developed based on collected videos [244]. Since the feasibility and cost of equipment for recording visual attention may hinder collecting data from navigators on either real

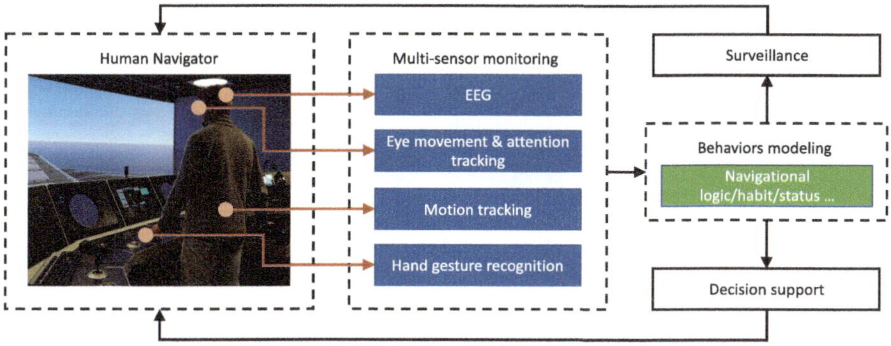

Fig. 6.5 Intelligent ship-bridge system for research on HITL navigational operations

ships or simulators, a low-cost and non-intrusive solution is demanded. By developing an integrated system that consists of a consumer camera and a deep-learning algorithm, it can trace navigators' visual attention at a high frequency (depending on the frame frequency of cameras) within specific visual attention zones (VAZ, also known as areas of interest (AOI)).

6.2.2 *Operation Response Data from Experiment Platforms*

The simulators in Fig. 6.6a and b are from Kongsberg, and the former is for research aim while the latter is for navigator training. The ship-bridge system with fixed seat shown in Fig. 6.6c is often seen on compact vessels and catamarans, and this

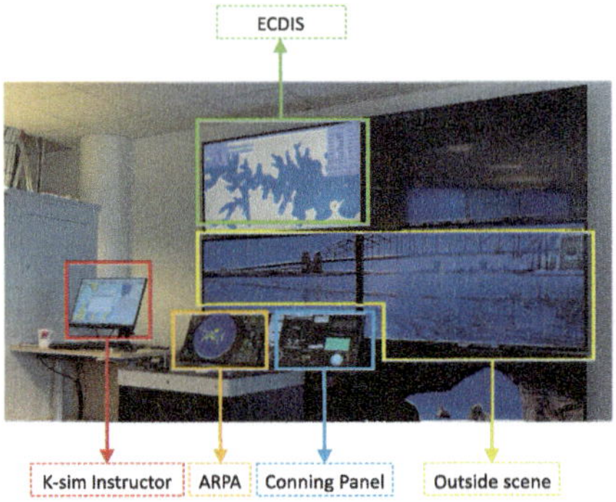

(a) Compact simulator from Kongsberg for preliminary research design.

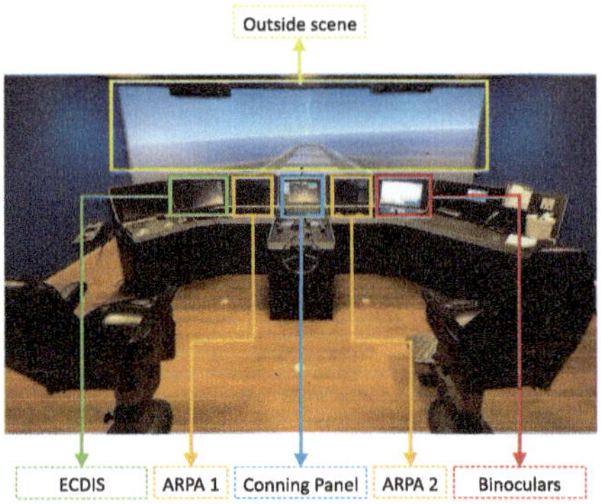

(b) Kongsberg maritime training simulator.

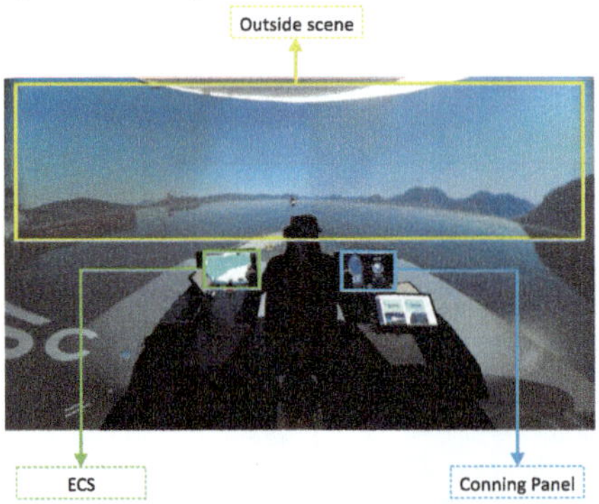

(c) Immersive ship-bridge simulator powered by Off-shore Simulator Centre AS (OSC).

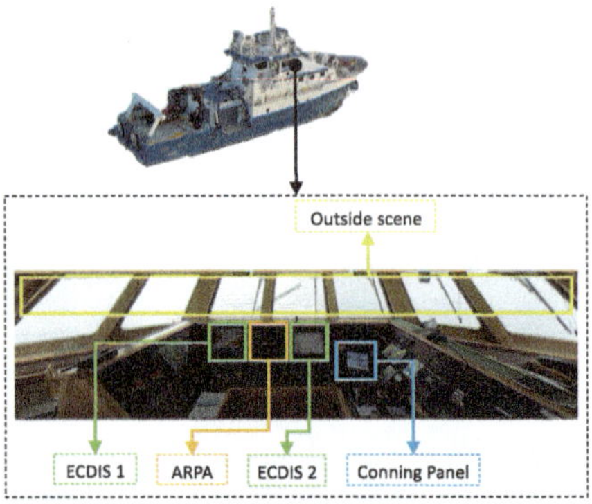

(d) NTNU's research vessel R/V Gunnerus and its ship bridge.

Fig. 6.6 Experimental platforms. (Terminology: Electronic Chart Display and Information System (ECDIS); Electronic Chart Systems (ECS); Automatic Radar Plotting Aids (ARPA)) simulator projects the outside scene on a dome screen which provides an immersive navigational experience. Figure 6.6d is the ship bridge of our research vessel R/V Gunnerus. All these simulators and the research ship are used as the experimental platform for carrying out HITL research [12].

Data collected from these platforms can be sorted into two categories: ship motion response (including heading/course and motions in six degrees of freedom) and machinery actuators (including readings of thrusters and rudder). Please refer to Table 4.2 for details.

6.2.3 Data Processing and Analysis

As shown in Fig. 6.7, there are two different ways to deal with the collected data. One is data interpretation based on expertise, and the other is the algorithm-based data-driven method. The two methods have their virtues:

- Expertise interpretation: First, expertise may bring a qualitative assessment of the data and rationalize the logic of navigational operations; it is not handful for them to discover existing problems; a step further, by using analytical methods to calculate correlations between different variables, and conducting interviews with participants, expertise interpretation may give an intuitive and comprehensible result. This may prove a boon to navigational training and teaching.
- Data-driven algorithms: Mature machine learning algorithms such as SVM and neural network/deep learning have a satisfying performance in classification and pattern recognition tasks. They are qualified to track navigators' visual attention, body motion, and gestures. It is a good choice for developing automatic functions and automated products, but as the calculation is done as if from a black box, the processing is difficult to clarify and explain. Nevertheless, this still triggers the research interest in explainable artificial intelligence.

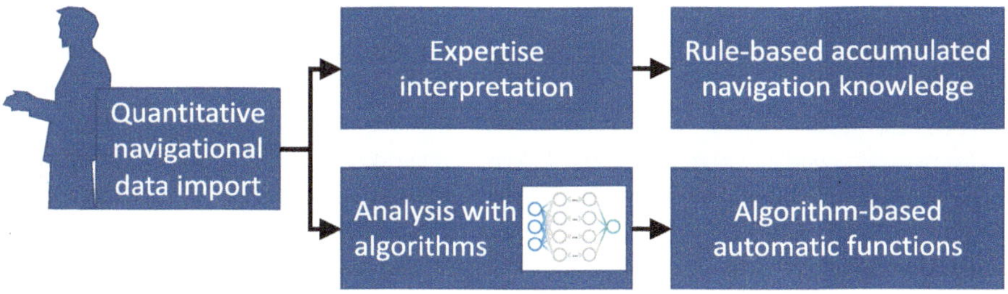

Fig. 6.7 Analysis procedures

6.2.4 Human-Centered Decision Support

Some applied examples are shown in Fig. 6.8. In Fig. 6.8c, the wearable eye-tracker can present precise visual transmission paths and glance durations. As an alternative, Fig. 6.8a shows an attention tracking method by feeding a deep learning neural network with images of facial and head orientational information to train an algorithm to trace the navigator's visual attention (e.g., the navigator is looking at the green zone in Fig. 6.8a) [244]. Figure 6.8b illustrates an onboard motion tracking method by using machine vision.

Figure 6.9 depicts the results of Fig. 6.8a, which shows three situations when the navigator looks at different VAZs. The bottom-left of each subfigure in Fig. 6.9 shows that eye-tracker glasses can locate the gaze to an exact point, although they fail to track eye movement in some cases. The algorithm can also recognize different VAZs, which means an approximate area of visual attention. Figure 6.9 proves the accuracy of the algorithm in the trial sailings in the designed scenarios.

From the above illustrations, these applications provide the chance to passively monitor navigators' operations. With further developed algorithms, when anomalies in ship response data and human-related data occur, human-centered decision support can effectively issue alerts or provide early warning before the navigation situation deteriorates.

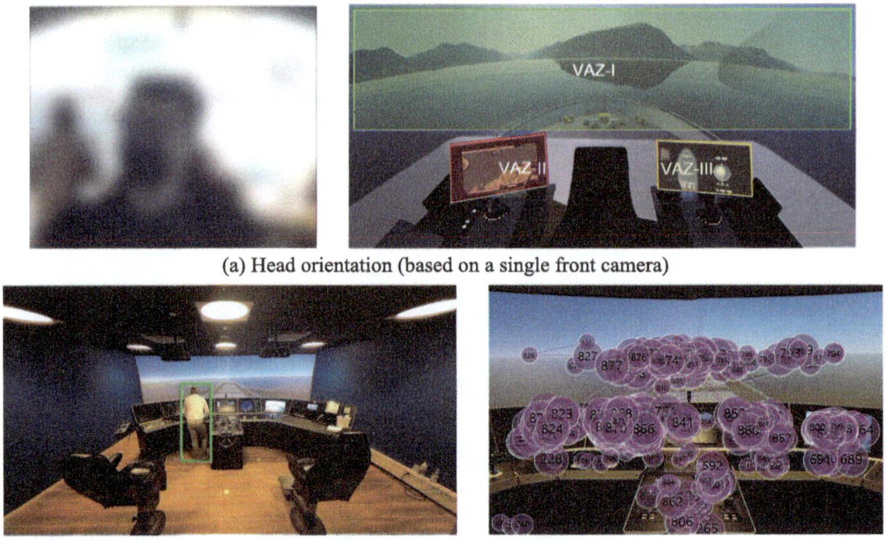

(a) Head orientation (based on a single front camera)

(b) Motion tracking (based on a single back camera)

(c) Gaze path by a wearable eye tracker

Fig. 6.8 Applied multi-sensor monitoring scenes

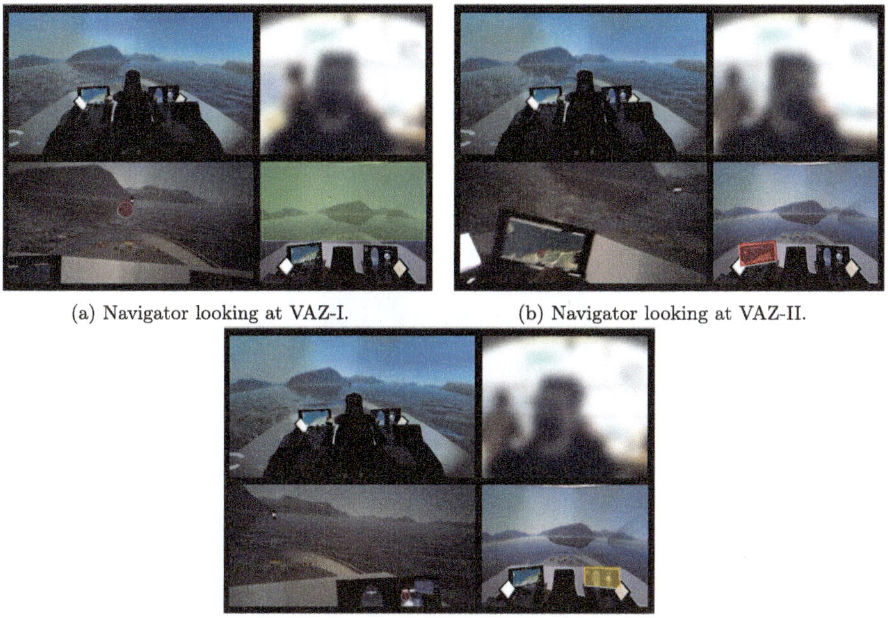

(a) Navigator looking at VAZ-I.

(b) Navigator looking at VAZ-II.

(c) Navigator looking at VAZ-III.

Fig. 6.9 Results plot of (1) top-left: back-view (only used for demonstration, not relevant to developing the algorithm); (2) top-right: front-view (input image to the algorithm; the facial image shown here is blurred according to GDPR); (3) bottom-left: eye-tracker glasses marked video; (4) bottom-right: the algorithm recognized VAZ is highlighted

6.3 Maritime Operation Support

Ideally, the digital twin provides the feasibility of dynamic autonomy among two extremes, i.e., the ship can be remotely operated by humans via either the land based communication network or the high bandwidth satellite communication systems, depending on the availability, quality and price of the autonomous ship; or the ship can be operated on its own pace using onboard sensor information.

Figure 6.10 proposes a solution for maritime operation support based on the digital twin enabling tools. It takes advantage of all digital information available for an asset including system and data information models, 3D visualization models, mathematical models, dependability models, condition and performance indicators and data analytics, to provide decision support, such as visualizing complex surroundings and illustrating future prediction of the situation for either achieving autonomy or remote control. Since the digital-twin system, e.g., a sea state estimation component, could be evaluated and updated via streamed sensor data and enabling tools if it is found there is a discrepancy between the estimation and the ground truth, the operational support tools thus could be able to react in various environmental conditions. These supporting tools will serve the industry for improving operational

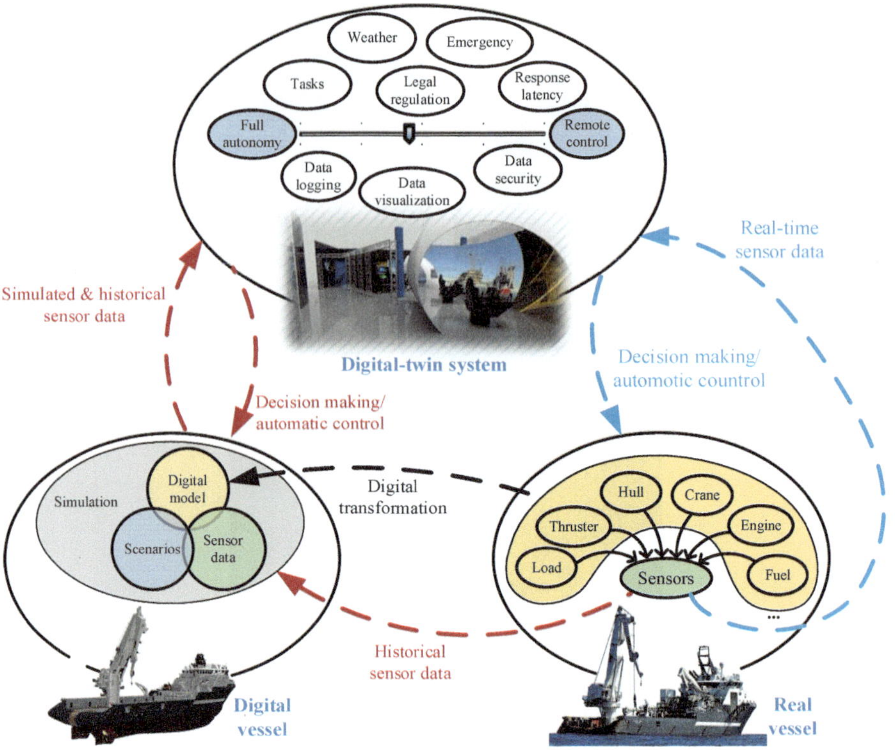

Fig. 6.10 A plausible digital-twin ship solution for maritime operation

effectiveness and safety. It is noted that they could also be used in digital twin-based simulation, indicating a potential use of these tools for maritime operation training.

6.4 Summary

Enabling tools serve as the middle-ware of the digital twin system, bridging the data source and the high-level applications. This chapter has introduced enabling technologies from two aspects. One is the data aspect, and the other is the human factor aspect. The former deals with raw data from data stream to dataset, while the latter concerns the role of human in use of high-level applications of the digital twin system. Related technologies and use cases are illustrated in this chapter. It is promising that by effectively combining these enabling tools mentioned above, one can develop efficient onboard support tools for either assisting the operator during marine operations or achieving automatic control of the operation. We will demonstrate it in Chaps. 7, 8 and 9.

Chunlin Wang received the Ph.D. degree from the Department of Ocean Operations and Civil Engineering, Norwegian University of Science and Technology (NTNU), Ålesund, Norway in 2023. His research interests include ship behavior analysis, sensitivity analysis, machine learning, and deep learning.

Baiheng Wu received the Ph.D. from the Norwegian University of Science and Technology (NTNU) in 2022. He is currently a post-doctoral research fellow with the Department of ICT and Natural Science at NTNU, Ålesund, Norway. His research interests include human factors, behavior analysis, human-in-the-loop for autonomous navigation.

Guoyuan Li received the Ph.D. degree in computer science from the Department of Informatics, Institute of Technical Aspects of Multimodal Systems, University of Hamburg, Hamburg, Germany, in 2013. In 2014, he joined the Department of Ocean Operations and Civil Engineering, Intelligent Systems Laboratory, Norwegian University of Science and Technology (NTNU), Ålesund, Norway, where he is currently a Professor in ship intelligence. He has published more than 150 articles in the areas of his research interests, including the modeling and simulation of ship motion, autonomous navigation, intelligent control, optimization algorithms, and the locomotion control of bioinspired robots.

Hans Petter Hildre is professor and head of the Department of Ocean Operations and Civil Engineering at the Norwegian University of Science and Technology (NTNU). His area of interest is product design and system architecture design. Hans Petter is the director for Centre for Research Driven Innovation (SFI-MOVE) within marine operations. This is cooperation between NTNU, SINTEF, University Sao Paulo and 15 companies at the west coast of Norway. Professor Hildre is head of research in national program Global Centre of Expertise Blue Maritime, project leaders in several research projects, member of the board in 5 companies, and has a number of patents.

Houxiang Zhang received the Ph.D. degree in mechanical and electronic engineering from Beihang University in 2003 and the Habilitation in informatics from the University of Hamburg, Germany in 2011. He is currently a full Professor with the Department of Ocean Operations and Civil Engineering, Faculty of Engineering, Norwegian University of Science and Technology (NTNU), Norway. From 2011 to 2016, he also hold a Norwegian National GIFT Professorship on product and system design funded by the Norwegian Maritime Centre of Expertise. He has engaged into two main research areas, including control, optimization and AI application especially on autonomous vehicle; and marine automation, digitalization and ship intelligence. He has applied for and coordinated more than 30 projects supported by Norwegian Research Council (NFR), German Research Council (DFG), EU, and industry. In these areas, he has published over 300 journals and conference papers as the author or coauthor. He was nominated and selected as a member of the Norwegian Academy of Technological Sciences in 2019, and a member of Academy of the Royal Norwegian Society of Sciences and Letters in 2024.

Chapter 7
Ship Trajectory Prediction for Decision Support

Tongtong Wang, Robert Skulstad, Lars Ivar Hatledal, Guoyuan Li, and Houxiang Zhang

Ship motion prediction considers constructing a model of vessel motion from historical data or its own dynamic characteristics and predicting its future trajectory. Since vessel motion prediction methods vary depending on their future coverage, short-term and long-term vessel motion predictions are exploited separately.

Predicting the motion of ships can be done in several ways, depending on the amount of information available about the specific ship. If the speed over ground (SOG) and course over ground (COG) is known, which is often the case when sensing approaching ships via radar or receiving Automatic Identification System (AIS) data, the trajectory may be derived by means of extrapolation of course and speed. This approach performs well when a ship maintains a speed and course for longer periods of time, for instance, when transiting in open waters or along common shipping lanes.

However, when the ship operator or autopilot performs course changes, the extrapolation of SOG and COG will produce significant trajectory prediction errors. For this case, a model that is able to reflect the dynamic properties of the ship will produce a more sensible result.

7.1 Prediction Models

Of the various methods to perform the short-term prediction of ship trajectory, we generally have two distinct categories of models:

- **Physics-based maneuvering models:** specific physical phenomena are modelled to reflect the dynamic behavior of the ship given input factors such as the ship velocities, control commands and environmental conditions.
- **Machine learning dynamics models:** based on no specific structure (subject to manual or automatic hyperparameter optimization), this model type applies data sampled onboard a ship to optimize the parameters of a dynamics model. This

H. Zhang et al., *Digital Twins for Vessel Life Cycle Service*,
https://doi.org/10.1007/978-981-97-4768-9_7

model is usually a machine learning model that reflects partial ship dynamics, such as the translational velocities.

We'll elaborate on these two modelling approaches in the following sections.

7.1.1 Physics-Based Maneuvering Model

The physics-based models describing ship dynamics could be formulated in the frequency or time domains. When the model structure is determined, proper parameters must be tuned within physically reasonable ranges to reflect the specific properties as close to reality. In this section, the physics model of the ship motion will be elaborated.

The 3 degree of freedom (DOF) horizontal plane maneuvering model describes relations between actuators, external environmental disturbances, and the ship hull [245]. The ship model based on the rigid-body kinetics is expressed as (7.1):

$$\begin{aligned} &\dot{\eta} = R(\psi)\nu \\ &M_{RB}\dot{\nu} + C_{RB}(\nu)\nu + M_A\dot{\nu}_r + C_A(\nu_r)\nu_r + D(\nu_r) + D_n(\nu_r)\nu_r = \tau_c + \tau_{wind} + \tau_{wave} \end{aligned} \tag{7.1}$$

where $\eta = [x, y, \psi]^T$ is the ship position vector containing the north, east positions and yaw angle in the Earth-tangential North-East-Down (NED) frame. $\nu = [u, v, r]^T$ is the ship velocity vector in surge (longitudinal axis), sway (lateral axis), and yaw (rotation about the up-down axis) directions in the ship's body coordinate, respectively. $R(\psi)$ is the horizontal plane rotation matrix which is given as (7.2):

$$R(\psi) = \begin{bmatrix} cos(\psi) & -sin(\psi) & 0 \\ sin(\psi) & cos(\psi) & 0 \\ 0 & 0 & 1 \end{bmatrix} \tag{7.2}$$

In the ship's coordinate frame, the model reflects the ship response due to the forces including wind, waves, thrusters, hull friction, and inertia. Here, $\nu_r = \nu - \nu_c$ is the relative velocity vector through water, $\nu_c = [u_c, v_c, 0]^T$ is the generalized ocean current velocity of an irrotational fluid. $M_{RB} \in \mathbb{R}^{3\times3}$ is the vessel's rigid-body mass matrix, and M_A is the added mass matrix. $C_{RB}(\nu) \in \mathbb{R}^{3\times3}$ and $C_A(\nu_r) \in \mathbb{R}^{3\times3}$ are the skew-symmetric Coriolis and centripetal matrices of the rigid body and the added mass. $D(\nu_r) \in \mathbb{R}^{3\times3}$ and $D_n(\nu_r) \in \mathbb{R}^{3\times3}$ are the linear, and nonlinear damping matrices, which are functions of the relative velocity ν_r between the vessel and the current.

External force components generally come from three sources (the ocean current-induced force has been accounted by the relative velocity ν_r):

- $\tau_c \in \mathbb{R}^3$— the propulsion force vector consisting of forces and moments produced by the thruster system;
- $\tau_{wind} \in \mathbb{R}^3$— force induced by wind acting at a certain angle relative to the ship;

- $\tau_{wave} \in \mathbb{R}^3$— force induced by waves. This force is not directly measurable and the drift component is generally observed as a nonzero slowly varying process.

The generalized propulsion force τ_c is obtained by translating the individual thruster forces, that act at specific locations on the ship hull, to the ship's body frame, as written in (7.3).

$$\tau_c = \begin{bmatrix} \tau_x \\ \tau_y \\ \tau_n \end{bmatrix} = T(\delta)F_T \tag{7.3}$$

where δ is the thruster orientation angle, and $T(\delta)$ is the thrust configuration matrix, which describes the geometrical locations of the thrusters. F_T consists of forces produced by tunnel thrusters, main propellers, or azimuth thrusters, respectively. The individual propeller thrust T and torque Q are generally formulated as a function of shaft speed n in revolution-per-minute, time-varying states $\mathbf{x}_p$, and fixed thruster parameters θ_p [246]. The thruster models (7.4) are generic models parameterized to fit the specific vessel.

$$\begin{aligned} T &= f_T(n, \mathbf{x}_p, \theta_p) \\ Q &= f_Q(n, \mathbf{x}_p, \theta_p) \end{aligned} \tag{7.4}$$

The wind force is the only environmental disturbance that can be estimated based on direct measurements. The deterministic model to estimate wind forces is given in (7.5).

$$\tau_w = \frac{1}{2}\rho_a V_{rw}^2 \begin{bmatrix} C_X(\gamma_{rw})A_{FW} \\ C_Y(\gamma_{rw})A_{LW} \\ C_N(\gamma_{rw})A_{LW}L_{oa} \end{bmatrix} \tag{7.5}$$

The relative wind speed is defined as $V_{rw} = \sqrt{u_{rw}^2 + v_{rw}^2}$ and attack angle $\gamma_{rw} = -atan2(v_{rw}, u_{rw})$, where $u_{rw} = u - V_w \cos(\beta_w - \psi)$, and $v_{rw} = v - V_w \sin(\beta_w - \psi)$. V_w and β_w represent the wind speed and its direction, respectively. C_X,C_Y, and C_N are wind coefficients specific for the hull or superstructure shape. A_{FW} and A_{LW} are frontal and lateral projected areas and L_{oa} is the overall length of the ship.

7.1.2 Machine Learning Dynamics Model

Instead of imposing the equations of motions for the horizontal plane motion of the ship as described in Sect. 7.1.1, a general machine learning model may be applied. Figure 7.1 shows an example of how this may look.

On the left of the figure an input vector, X_t, contains measurements from the ship taken at time t with a sampling interval of Δt. In principle, measurements from any sensor can be used. However, the most relevant measurements, also known as features when used in machine learning applications, are often selected based on either feature

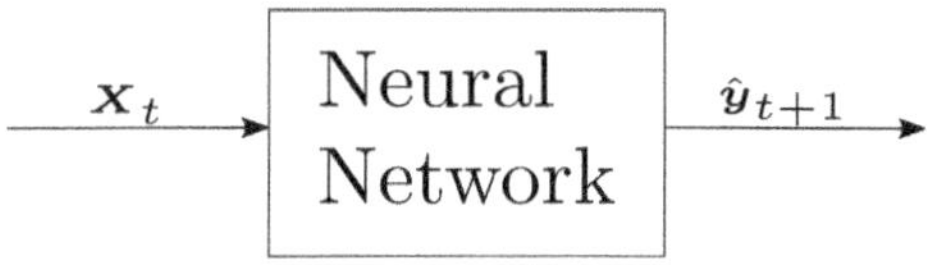

Fig. 7.1 A general machine learning prediction model. In this case it predicts one time step into the future

selection methods or knowledge of ship hydrodynamics [247, 248]. $\hat{\boldsymbol{y}}_{t+1}$ refers to the predicted ship state at the next discrete time step (i.e. $\Delta t = 1s$). Depending on the prediction strategy, this may be the velocities [249] or position [156].

Assuming that the sample interval is 1 s, making a one-step prediction of the ship trajectory is not very useful. The large inertia of ships suggests that longer prediction horizons should be used, such that ship operators may apply this information in their situational awareness assessment, or plan course corrections to e.g. avoid nearby obstacles.

Various strategies for increasing the prediction horizon have been applied within time series prediction [250]. We'll consider two fundamentally different approaches here:

- Iterative: the model shown in Fig. 7.1 is iterated and the input, $\boldsymbol{X}_t$, is updated with predicted values from $\hat{\boldsymbol{y}}_{t+1}$ as the prediction time advances. This can be realized in a nonlinear autoregressive exogenous (NARX) network or in recurrent networks that inherently manage this flow of temporal information. An autoregressive method for ship motion prediction is given in [251].
- Direct: a prediction of $\hat{\boldsymbol{y}}_{t+1,t+h} = [\hat{\boldsymbol{y}}_{t+1}, \ldots, \hat{\boldsymbol{y}}_{t+h}]$ is made directly by using information available from the time of prediction [252].

These two methods have some clear advantages and disadvantages [253]. The direct approach does not face the issues related to the accumulation of errors. However, it typically has larger input/output dimensions leading to more complex models. Comparatively, the iterative approach is intuitive and has lower complexity.

The process of obtaining a machine learning model of ship dynamics is similar in many ways to acquiring parameters of the physics-based equations of motion (EOM) that is prevalent within the motion control community [254]. A major difference is the fixed structure of the EOMs and a strict regime for generating the model-scale data for the physics-based model. The graphic below Figure 7.2 shows the two approaches for generating a model that represents the dynamics of a ship.

It is clear that more effort is required when establishing a physics-based model through system identification, compared with a ML model, for motion prediction. The modelling freedom—allowing designers to incorporate any measurement and model structures—gained from using ML methods is attractive. But the designer must also accept the challenges in using these models: the main challenge being the explainability of the predictive model. As the model structure in general does not support physically interpretable connections, and the number of parameters is usually much larger than models originating from physics, one is left with an un-interpretable model [256].

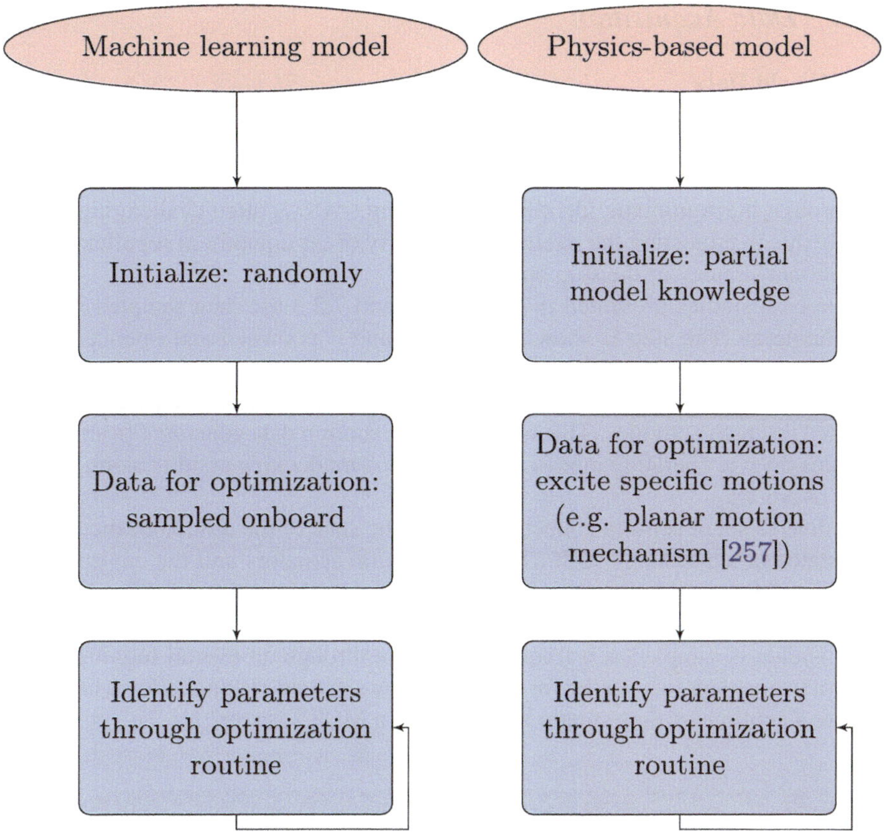

Fig. 7.2 A schematic view of the two separate model optimization schemes: machine learning and physics-based

This is the motivation for the domain of hybrid modelling. In this domain, knowledge about the motion of ships derived from physics is combined with models that learn from data sampled on the ship. In that sense, the system identification approach, mentioned above, can be categorized as a hybrid method given that partial knowledge is injected prior to optimizing the model parameters.

7.2 Case Studies

The case studies presented in this section highlight some of the modelling principles discussed in Sect. 7.1.

7.2.1 Data Acquisition

Real-World Data

All machine learning modelling approaches rely on the availability and quality of data. And procuring real-world datasets that go beyond the features of data transmitted through the Automatic Identification System (AIS) is often challenging. This is mainly due to protecting the intellectual property of the equipment suppliers and the operational strategy of the ship operator.

The case studies presented in Sects. 7.2.2 and 7.2.3 use data sampled from the RV Gunnerus. The ship is shown in Fig. 7.3 and it is owned and operated by the Norwegian University of Science and Technology (NTNU). The operation of the RV Gunnerus is related to both in-house educational and research activities as well as external research activities. This means that the motion data generated from the ship contains diverse operating modes as well as frequent docking at ports along the west coast of Norway.

A diverse set of sensors collect data about the state of the motion of the ship, but also states of subsystems such as the generators, actuators and the environmental condition. However, only sensors directly related to the motion of the ship will be introduced at this point (Table 7.1).

This data is sampled at a frequency of 1 Hz through an overall logging system that transmits data to a server for offline use. A separate channel allows users who require a continuous data stream to connect to a MQTT server.

Fig. 7.3 The RV Gunnerus in stationkeeping mode

Table 7.1 A subset of the data channels sampled onboard the R/V Gunnerus. The bottom two rows reflect the variables used in the data extraction procedure described in Sect. 7.2.2. The table is copied from [257]

Data channel	Unit
North position	m
East position	m
Heading angle	deg
Surge velocity	knots
Sway velocity	knots
Heading rate	deg/s
Roll angle	deg
Pitch angle	deg
Heave displacement	m
Roll rate	deg/s
Pitch rate	deg/s
Heave rate	m/s
Wind direction	deg
Wind speed	knots
Course angle	deg
Speed over ground	knots
Port thruster rotational speed	RPM
Port thruster angle	deg
Starboard thruster rotational speed	RPM
Starboard thruster angle	deg
Tunnel thruster rotational speed	%
Drive running	Boolean
Motor at zero speed	Boolean

Simulated Data

Although maritime ship simulators provide an idealized reality, where the user may specify all operational conditions, define disturbances and measure any simulated state with zero noise, they serve as platforms for the development of new functionality. As long as the user is aware of the discrepancies they may be accounted for in the process of developing new functionality.

The simulated data is applied in the case study presented in Sect. 7.2.4 and comes from experiments conducted in the commercial simulator software developed by Offshore Simulator Centre (OSC) AS. In this simulated environment, the wave, wind, and ocean current characteristics may be manipulated to induce a wide range of environmental loads on the ship. Figure 7.4 shows the virtual ship, which is based on the Olympic Zeus.[1] It is a multi-purpose offshore vessel with an overall length of 93.8 m and a deadweight of 4925 t. The physics-model of the ship is provided

[1] The Olympic Zeus ship, https://www.olympic.no/fleet--chartering/subsea--renewable/olympic-zeus/.

Fig. 7.4 A screenshot of the OSC simulator environment. The panel on the left allows the user to make changes in the external disturbances affecting the ship

by the FhSim[2] software platform created by SINTEF. In order to make the virtual measurements more realistic, noise was added to the north/east position and heading angle (sum of bias, white noise and a Gauss-Markov process) and velocities (bias and white noise) [258].

7.2.2 Demo of Harbor Docking

Here, we take harbor docking operation as the first example to illustrate the possible solutions using the digital twin system, as shown in Fig. 7.5. In recent years, the maritime industry including Kongsberg, Wärtsilä and DNV has put effort on implementing auto-docking systems. The challenges for this application lie in the uncertainties from ship model and environmental effects, and the nonlinear control.

Given the historical operation data and ship motion data, the OSP framework is applied to simulate ship dynamics. Then, we can build up docking scenarios in the simulator and invite ship masters to operate on different levels of sea conditions. The simulated data could then be fed into the DT to train the controller for ship docking, and the predictor for online estimation of ship motion. This process will be run in a short period if there is enough simulated data for training. Further validation can

[2] The FhSim software platform, https://fhsim.no/, Date accessed 29-June-2023.

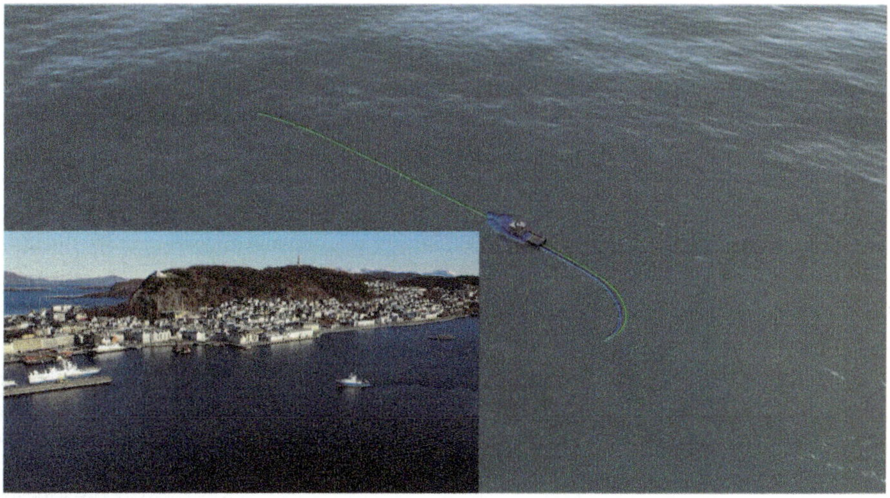

Fig. 7.5 Docking testing at Aalesund harbor in Norway. The grey ship and green trajectory present the real physical ship behavior; the blue ship and blue trajectory present the digital twin behavior; the video window shows the real operation as same as grey ship presented

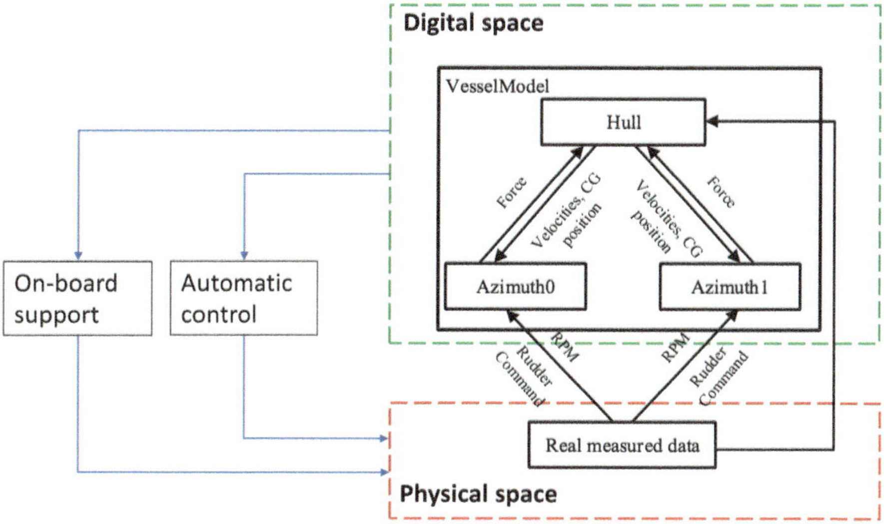

Fig. 7.6 The digital twin ship assists autodocking based on co-simulation

be achieved and compared between the docking operation in the field and digital twin simulation. If validated, the twin ship in the simulator can work in two modes, as shown in Fig. 7.6. The first mode is to provide onboard support for predicting ship motion. By feeding the deck-operation data and ship motion data into the DT, the predictive ship motion can be fed back to the ship master during the operation. The other mode is automatic control, which only takes ship motion data as input and produces operation commands for ship docking. In Fig. 7.6, each block in the "digital space" represents an FMU with the communication signal specified. The involved components are provided by different stakeholders.

A full-scale ship docking experiment is conducted in Aalesund harbour, Norway. When the ship Gunnerus is operated to dock to a berth, the implementation of twin ship docking in the simulator is going on simultaneously. By feeding it the in-field ship operation and environment sensor data, the twin ship produces highly resemble trajectories and speeds as shown in Figs. 7.7 and 7.8. While the twin ship is not a complete duplicate of the physical ship, we are seeking to make them as indistinguishable as possible. Slight mismatches are observed, especially in speeds. They might be caused by simulator fidelity, which is somewhat deteriorated by the unmodeled uncertainties. Otherwise, the discrepancies may result from instrumental errors, for instance, the wave and current measurements are usually difficult to accurately obtain. These unpredictable factors are unavoidable so far. Solutions to limit these factors will be further explored by trying their best to improve simulation fidelity and preprocess real-world measurements. Overall, the validation supports the twin ship fidelity and approves predictive and control applications.

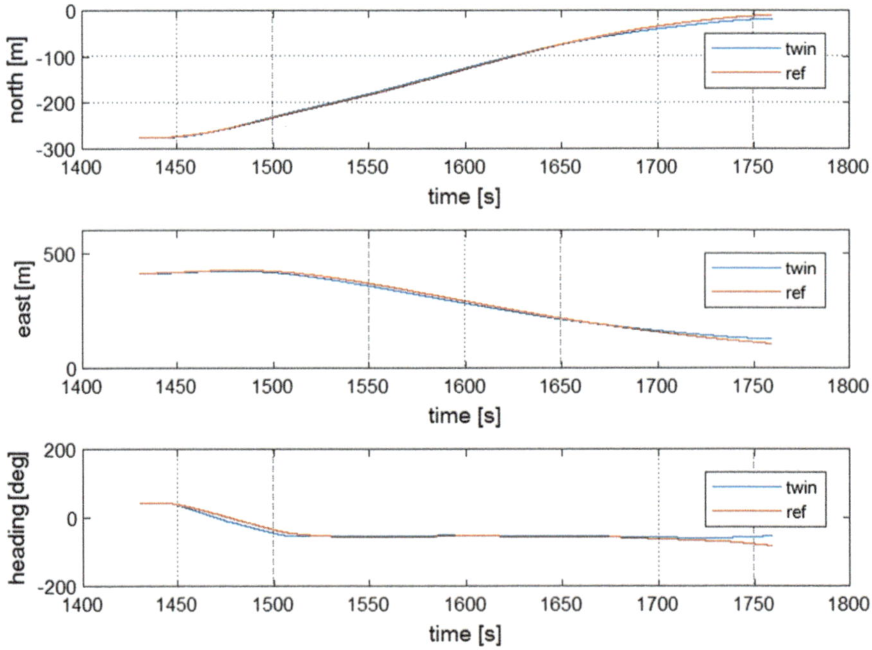

Fig. 7.7 The validation of twin-ship docking in a simulator versus real ship in the field in terms of positions. Ref: real ship experimental data

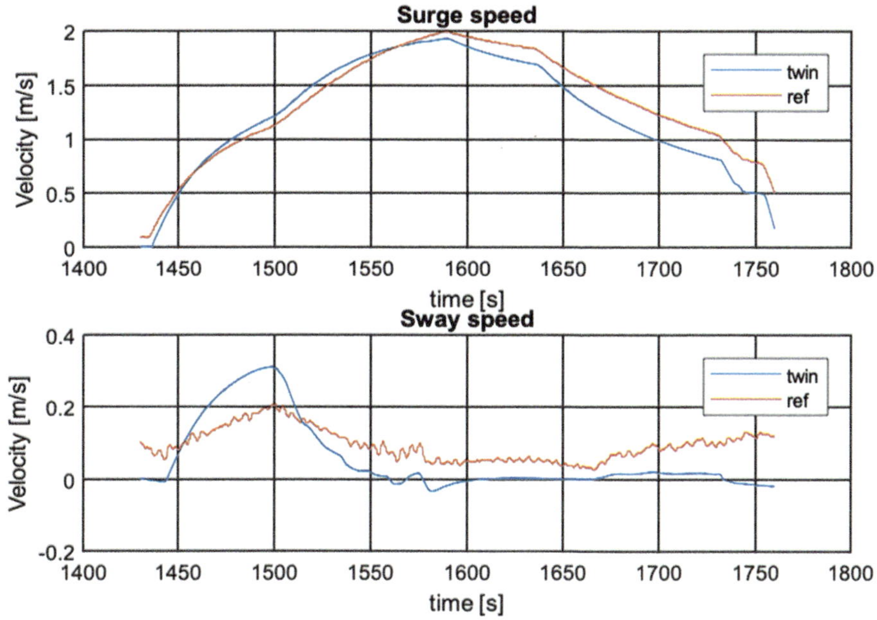

Fig. 7.8 The validation of twin-ship docking in a simulated ship versus the real ship in the field in terms of speeds

7.2.3 Motion Prediction During Zigzag Maneuver

The zigzag test is a standard maneuver used to compare the maneuvering properties and control characteristics of a ship with those of other ships. Definitions of zigzag can be referred to [259].

The hybrid cooperative model contains two complementary parts originating from two separate domains: physics-based and data-driven. The preliminary approximate ship dynamics is provided by the short vessel model, and the maneuvering data is sampled from the elongated ship zigzag experiments. The way they collaborate can be represented in Fig. 7.9.

The model groundwork (top yellow box) is built on a hydrodynamic model of a similar vessel. It serves to provide approximate ship states $\hat{X} = [\hat{\eta}_m, \hat{\nu}_m]^T$ reacting to the control command and environment configuration. $[RPM(t_0), \delta(t_0)]^T$ is the propulsion system feedback, and $[\beta_w(t_0), V_w(t_0)]^T$ refers to the wind conditions. Similar hydrodynamic properties characterize the reference model, and therefore, acceptable model dispersion is within expectation. Upon the preliminary model, the data-driven NN calibrator (bottom green box) is built to map the rough dynamics to a surrogate model that is able to accurately predict ship behaviors. It is believed that the preliminaries of the ship dynamics are carried forward into the data-driven model by means of informative input.

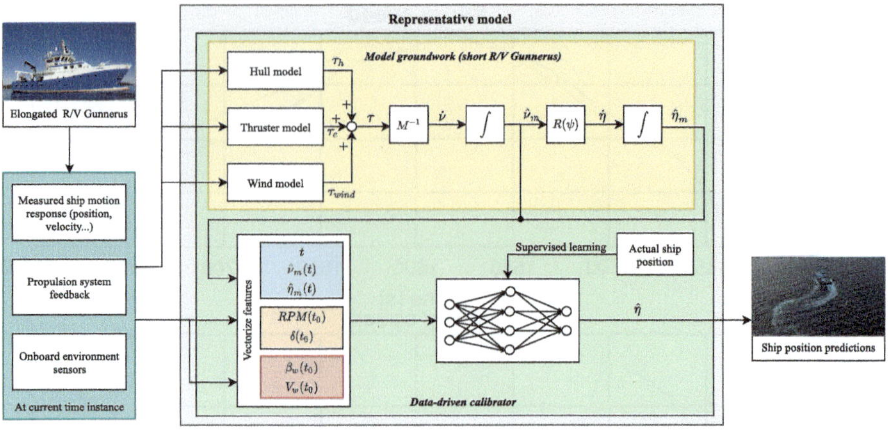

Fig. 7.9 A complete flowchart showing the incorporation of approximate dynamics into the data-driven calibrator

A fully connected feed-forward NN model is applied as a calibrator. Three hidden layers are specified in the network architecture, and each hidden layer contains ten neurons. The input features include prediction time ahead of the current instance, model predicted vessel velocities and positions in the horizontal plane, propulsion feedback, and external environmental factors. Supposing prediction starts at t_0, the corresponding input vector and desired output will be expressed as:

- Input: $[t_i, \hat{\nu}_m(t_0 + t_i | t_0), \hat{\eta}_m(t_0 + t_i | t_0), RPM(t_0), \delta(t_0), \beta_w(t_0), V_w(t_0)]$
- Output: $\hat{\eta}(t_0 + t_i | t_0)$

where $t_i \in [t_0, t_0 + t_h]$ is the forward time interval, $\hat{\eta}_m(t_0 + t_i | t_0)$ represents the reference model predicted positions starting from t_0. $RPM(t_0)$, $\delta(t_0)$, $\beta_w(t_0)$, and $V_w(t_0)$ are sensor data recorded at t_0.

Both simulation and full-scale experiments are implemented to validate the proposed hybrid approach.

Figure 7.10 shows the variation of average error of both approaches with respect to each training example. It indicates that the hybrid method works well on decreasing errors and improving prediction accuracy. In Fig. 7.11, the distributions of each error index are presented. On average, the hybrid predictor decreases the prediction errors to a significant extent, inducing a high-fidelity representative model of the elongated vessel. Particularly, a more noticeable calibration performance of y direction is observed compared to that of x direction.

The experiment results under different sea states are listed in Table 7.2. From the table, the hybrid modeling method shows a significant decrease in error predictions. A larger error reduction in the x position is addressed when the sea state gets severe. It is observed that the integration of the NN calibrator works well on improving the forecasting accuracy concerning various environments, demonstrating robustness.

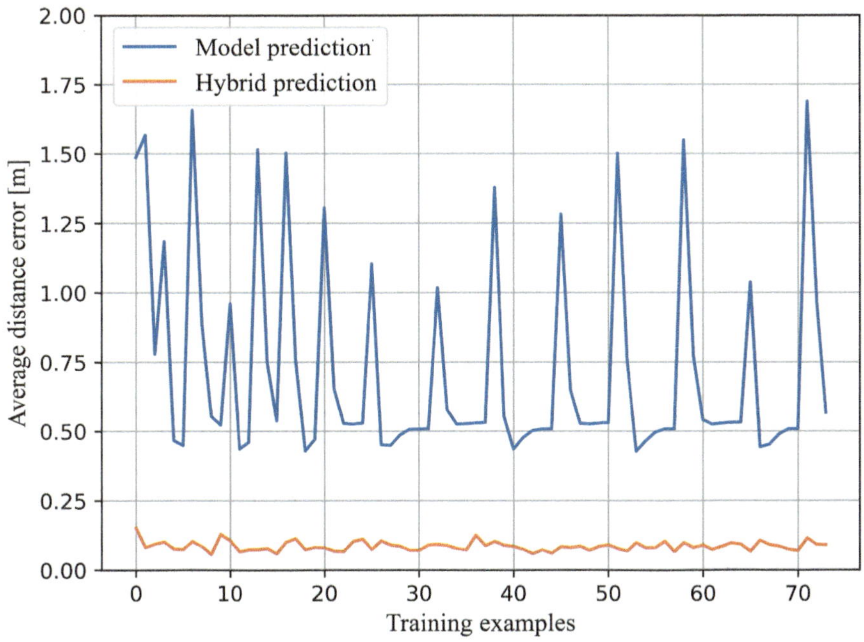

Fig. 7.10 The predictive performance of hybrid approach in simulation studies

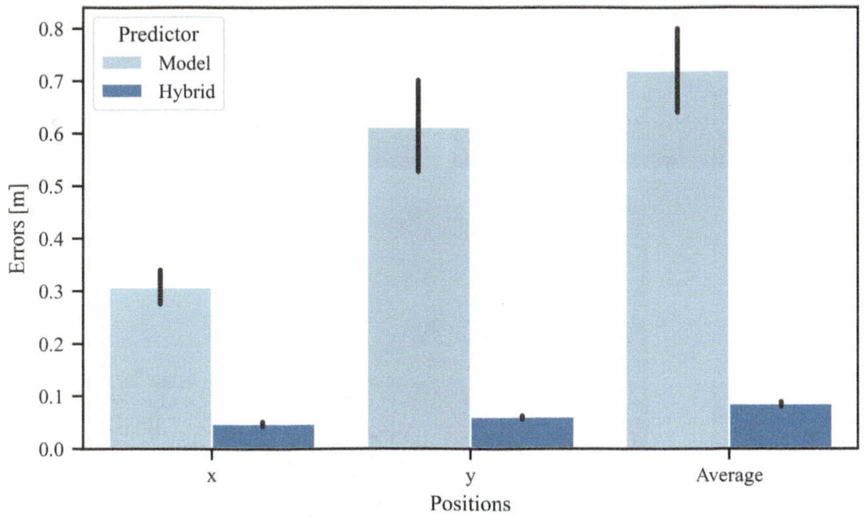

Fig. 7.11 The prediction errors of reference model and hybrid approach in different directions

The maneuvering experiment of elongated RV Gunnerus was conducted in November 2019 in Trondheim, Norway (see Fig. 7.12). During this process, thirteen sensor channels related to the ship motion of the vessel were sampled, including positions, velocities, environments, as well as commands.

Table 7.2 Evaluation of prediction errors of reference model and hybrid approach at different sea states

Sea states	Positions	Prediction errors		
		Model	Hybrid	Reduction (%)
Calm water	x	0.31	0.05	84.8
	y	0.61	0.06	90.3
	Average	0.72	0.08	88.3
Gentle	x	1.19	0.18	84.9
	y	1.61	0.06	96.3
	Average	2.19	0.20	91.0
Moderate	x	3.02	0.36	88.1
	y	1.56	0.10	93.6
	Average	3.60	0.39	89.2

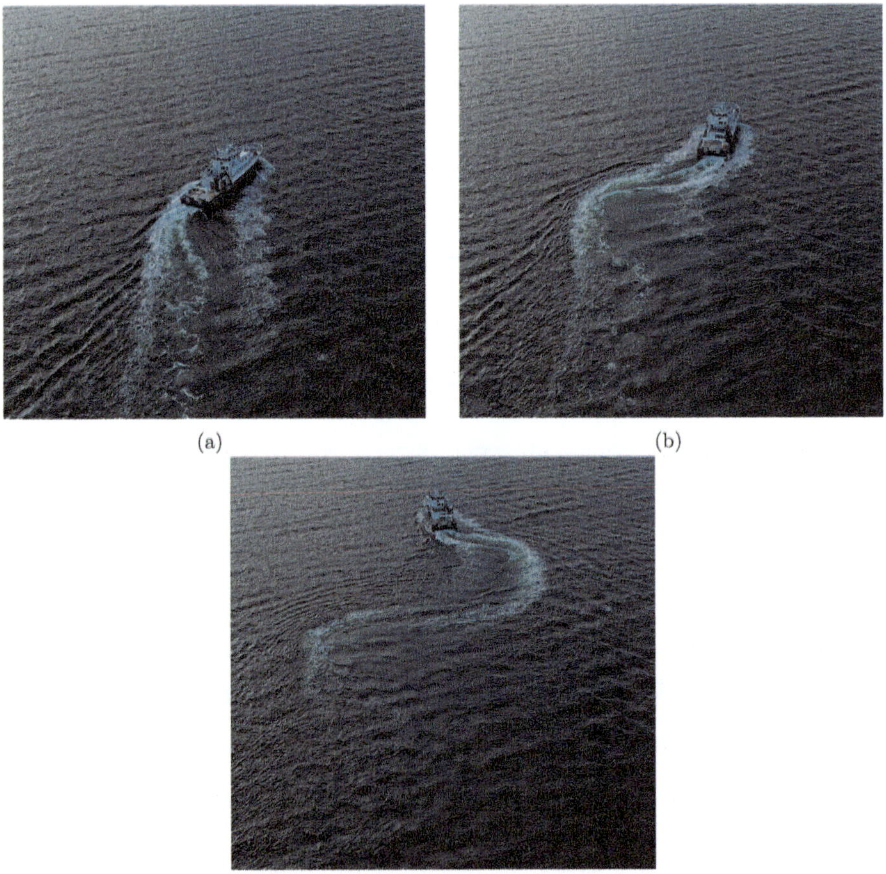

(a) (b) (c)

Fig. 7.12 RV Gunnerus when executing a zigzag maneuver on site in Trondheim, Norway

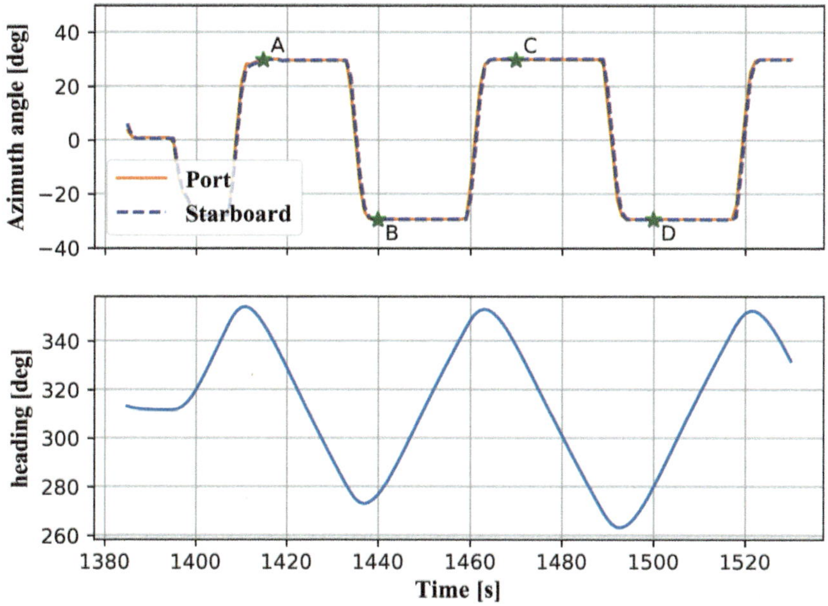

(a) Azimuth turning angle and ship heading during maneuvering.

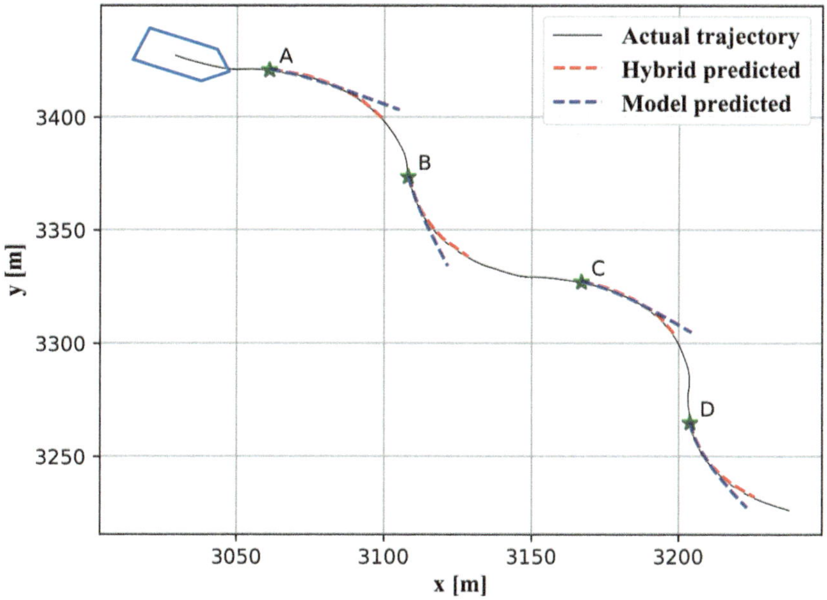

(b) Hybrid predictions in comparison with model predictions in real life.

Fig. 7.13 The predictive performance of hybrid approach in real-life experiment

During maneuvering, the thruster turning angle and ship heading are changing, as shown in Fig. 7.13a. In this process, the tunnel thruster is turned off. By integrating the sampled signals in the current instance as well as the preliminary mathematical model outputs into the neural network, the desired positions at the next instance are obtained. The prediction interval is 15 s, and the calibration results are verified as shown in Fig. 7.13b. The hybrid predictions have a satisfactory agreement with the actual ship trajectory compared with those propagated by the reference mathematical model. The hybrid predictive model is proven effective and can be applied in realistic ocean scenarios.

7.2.4 Dead Reckoning

If the ship suffers a loss of position references, dead reckoning may be used to infer the displacement of the ship throughout the outage [258]. In commercial maritime navigation solutions, this may be performed by model-based navigation systems that utilize the established dynamic model along with measurements of the actuators and the compass [259]. Inertial navigation systems may also perform dead reckoning using measurements from accelerometers, gyroscopes and compass [260]. These are normally aided by satellite positioning systems.

The aforementioned navigation systems utilize either a pre-determined vessel model or high-end inertial measurement units. The method proposed for dead reckoning in this section (see Fig. 7.14) is based purely on data measured from the ship from existing systems such as the thrusters (revolutions per minute/blade pitch angle),

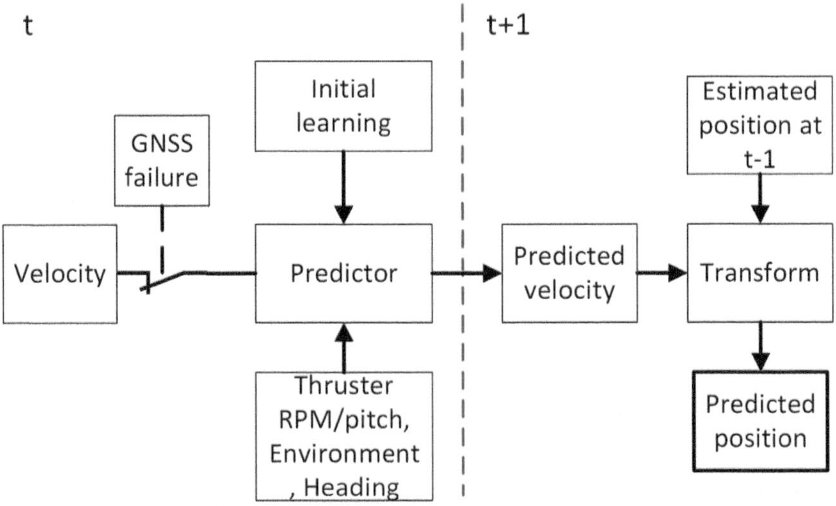

Fig. 7.14 The proposed modelling approach for dead reckoning during a satellite navigation outage

wind speed, wind direction and heading, and thus represents a more generic approach to dead reckoning. In the *Predictor*-block of Fig. 7.14 a long short-term memory (LSTM) recurrent neural network translates the input vector (actuator signals, wind measurements and compass heading) into velocities in the ship coordinate frame. These are subsequently transformed into positions in the North-East-Down frame through the rotation matrix and the measured compass heading.

The LSTM model mentioned above, is in reality two separate LSTM networks; one that predicts the next surge velocity and one that predicts the next sway velocity. This split allows the use of feature selection to optimize the input features for each predicted state. Hyperparameter optimization, of the learning rate and number of network nodes, further tailors each network to the specific modelling task.

The simulator and scenario described in Sect. 7.2.1 were used to test the method of dead reckoning. The simulated ship, equipped with two bow tunnel thrusters, two stern tunnel thrusters and two main propellers with rudder, was put in stationkeeping mode and exposed to a repeating sequence of increasingly severe sea states (see Table 7.3).

Each weather condition, i.e. a specific wave height and wind velocity at a certain direction, is simulated for 14 min. The first seven minutes of this interval is spent acclimatizing to the new external disturbances. 60 s towards the end of the 14-minute interval is used for running stationkeeping in dead reckoning mode. Figure 7.15 shows the results of running the proposed method along with two reference methods: a regular feedforward neural network in Fig. 7.15b and the Kalman filter approach that includes a linearized DP model of the ship in Fig. 7.15c.

The position estimation error for each sea state and direction, which is visualized in the polar plots as the radial distance at 30 degree increments, is determined by (7.6).

$$\bar{e}_{dist}[k] = \frac{1}{N}\sum_{k=1}^{N}\sqrt{(\hat{p}_n[k] - p_n[k])^2 + (\hat{p}_e[k] - p_e[k])^2} \tag{7.6}$$

(p_n, p_e) is the north and east position of the ship, $(\hat{p}_n, \hat{p}_e)$ is the corresponding dead-reckoned positions, and k is the evaluation time instance for each DR period. N equals the DR evaluation length of 60 s for each sea state.

Visualizing the errors calculated by (7.6) highlights the consistency and accuracy of the LSTM DR model relative to the feedforward model and Kalman filter model. The feedforward model provides similar consistency at a reduced accuracy (see Fig. 7.15b) while the Kalman filter model is less consistent, but has good accuracy

Table 7.3 The parameters of the sea states simulated at each discrete weather direction, [0, 30, ..., 330] degrees

Significant wave height (Hs) (m)	Wind velocity (ms^{-1})
1	2
2	4
3	7
4	11

Fig. 7.15 Results of mean position estimation error given in the horizontal plane for the LSTM, SLFN and KF model. Each data point shows the mean position estimation error during a one-minute DR period. Hs denotes the significant wave height in meters and is related to the sea states described in Table 7.3

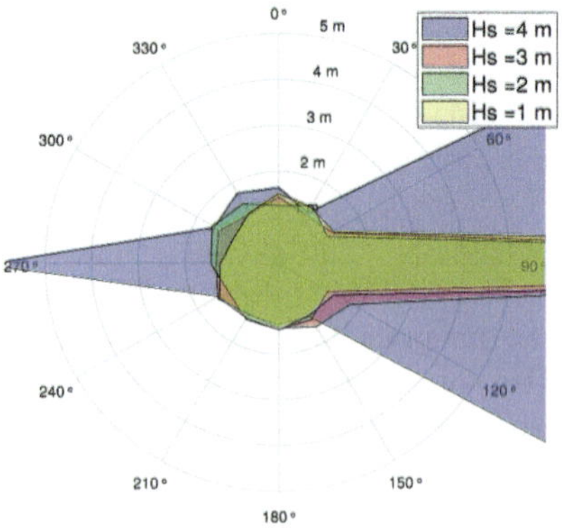

(a) LSTM

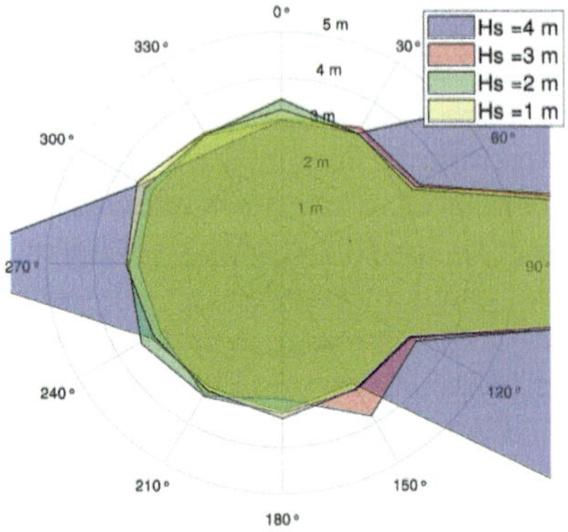

(b) SLFN

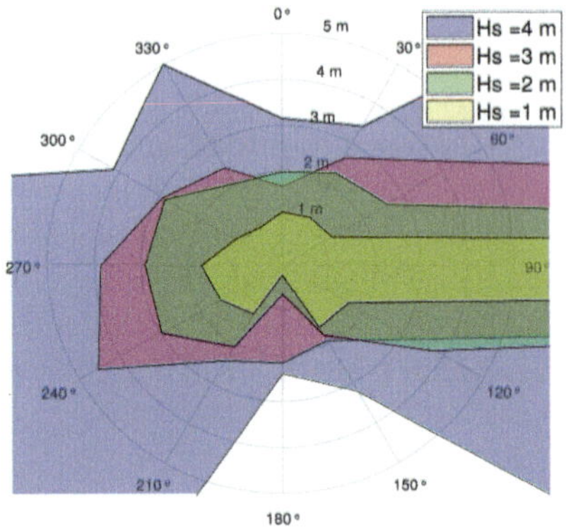

(c) KF

for less severe sea states. The large errors observed for wave heights of 4 m are because of saturation in the capabilities of the combined control and navigation (dead-reckoning) system due to large environmental forces.

7.2.5 Discussion

Over the years, physics-based methods, such as the Abkowitz and Maneuvering Modeling Group (MMG) models, have been used to describe ship motion dynamics and their consequences based on classic ship theory. In these methods, the hull-propeller-rudder interaction coefficients and hydrodynamic derivatives are regarded as critical parameters that require high attention to be determined with external tools, such as the captive model test, computational fluid dynamics, as well as system identification approaches. These tools are extremely useful for ship manoeuvring simulations and predictions at the ship design stage. However, they fail to present a unified and convincing prediction of ship motions under environmental conditions in real-time. In dealing with the uncertainty and randomness of environmental disturbances, unknown ship motion behavior characteristics, and unmodelled dynamics, intelligent data-driven algorithms play a significant role.

However, the data availability in a form suitable for analysis limits the advance of machine learning in marine science. The amount of annotations can be constrained and the quality variable and it is too expensive to get more and/or better data.

In fact, except for pursuing prediction accuracy which has been the long-lasting and sole standard for comparing the performance of different models, robustness and reliability should also be highlighted, as trust becomes very important when the output from these systems are intended as input to models for estimation or robust control. Recent studies have shown that modern deep neural network models can be fooled by slightly perturbed inputs, known as adversarial examples, and they would cause neural networks to malfunction. The trade-offs between robustness and accuracy should be assessed when training a data-driven model.

Although extensive different case studies have been implemented, the answer to the question of selecting the "best model" is still pending. Being aware of the strengths and drawbacks of each predictive modeling discipline, choices are subject to the task requirement.

7.3 Summary

Ship motion prediction is helpful for various naval operations, such as aircraft landing, cargo transfer, off-loading of small boats, and ship mating between a big transport ship and some small ships. Five to thirty seconds ahead prediction can give the operator increased awareness of his/her surroundings. Improving the accuracy of ship motion prediction is conducive to decision-making for performing motion sensitive maritime activities and avoiding potential navigation risks. Predictions can be achieved by physical models based on the ship's hydrodynamic coefficients and equations, intelligent models based on time series analysis and intelligent algorithms, and physics-data hybrid models. A proper model should be formulated according to its engineering requirements.

Tongtong Wang received her Ph.D. degree in Engineering from the Norwegian University of Science and Technology (NTNU), Alesund, Norway, in 2022. She is currently working as a researcher with the Department of Ocean Operations and Civil Engineering, NTNU. Her current research interests include physics-informed machine learning and its application in marine navigation and operation.

Robert Skulstad received the Ph.D. degree from the Department of Ocean Operations and Civil Engineering, Norwegian University of Science and Technology (NTNU), Alesund, Norway, in 2021. He is currently a Post-Doctoral Fellow with the Department of Ocean Operations and Civil Engineering, NTNU. His research interests include the use of machine learning for motion prediction and control in maritime operations.

Lars Ivar Hatledal received the Ph.D. degree from the Department of Ocean Operations and Civil Engineering, Norwegian University of Science and Technology (NTNU), Alesund, Norway, in 2021. He is currently working as an Associate Professor with the Department of ICT and Natural Sciences. His research interests revolves around software architecture, co-simulation and visualisation.

Guoyuan Li received the Ph.D. degree in computer science from the Department of Informatics, Institute of Technical Aspects of Multimodal Systems, University of Hamburg, Hamburg, Germany, in 2013. In 2014, he joined the Department of Ocean Operations and Civil Engineering, Intelligent Systems Laboratory, Norwegian University of Science and Technology (NTNU), Alesund, Norway, where he is currently a Professor in ship intelligence. He has published more than 150 articles in the areas of his research interests, including the modeling and simulation of ship motion, autonomous navigation, intelligent control, optimization algorithms, and the locomotion control of bioinspired robots.

Houxiang Zhang received the Ph.D. degree in mechanical and electronic engineering from Beihang University in 2003 and the Habilitation in informatics from the University of Hamburg, Germany in 2011. He is currently a full Professor with the Department of Ocean Operations and Civil Engineering, Faculty of Engineering, Norwegian University of Science and Technology (NTNU), Norway. From 2011 to 2016, he also hold a Norwegian National GIFT Professorship on product and system design funded by the Norwegian Maritime Centre of Expertise. He has engaged into two main research areas, including control, optimization and AI application especially on autonomous vehicle; and marine automation, digitalization and ship intelligence. He has applied for and coordinated more than 30 projects supported by Norwegian Research Council (NFR), German Research Council (DFG), EU, and industry. In these areas, he has published over 300 journals and conference papers as the author or coauthor. He was nominated and selected as a member of the Norwegian Academy of Technological Sciences in 2019, and a member of Academy of the Royal Norwegian Society of Sciences and Letters in 2024.

Chapter 8
Prognostics and Health Management (PHM) System

Peihua Han, André Listou Ellefsen, Vilmar Æsøy, and Houxiang Zhang

8.1 Introduction to Prognostics and Health Management (PHM) System

Maintenance in shipping today follows either a reactive maintenance (RM) or preventive maintenance (PM) approach [261]. RM introduces high risks of unscheduled downtime, while PM provides relatively high reliability, but at unnecessary costs due to predetermined maintenance intervals. PM also will not detect random failures which are in fact the most common failure pattern in the maritime industry. Thus, a more predictive maintenance approach is necessary to identify these kinds of failures. A predictive system will significantly increase operating performance and drastically decrease unexpected system failures [262].

Over the past decade, the field of Prognostics and Health Management (PHM) has emerged as a promising discipline in engineering, offering valuable support for predictive maintenance decision-making. In particular, PHM has been successfully integrated by entities such as the US Department of Defense [263], the aerospace industry [264], and the aviation industry [265]. PHM strives to decrease and ultimately eliminate inspections and time-based maintenance intervals. This will be achieved through accurate condition monitoring (CM), precursor and/or incipient fault detection, isolation and identification, and prediction of approaching failures. However, in the maritime industry, PHM has only recently begun to receive attention. The implementation of a comprehensive PHM system is crucial for ship autonomy, as it possesses advanced capabilities to detect, isolate, and identify early-stage or developing anomalies and faults in ship components and subcomponents. Additionally, it monitors and predicts the progression of such anomalies and faults, while providing decision support or automation to optimize maintenance schedules [266]. This becomes particularly significant for ship autonomy, where there is little to no crew involvement onboard the vessel.

H. Zhang et al., *Digital Twins for Vessel Life Cycle Service*,
https://doi.org/10.1007/978-981-97-4768-9_8

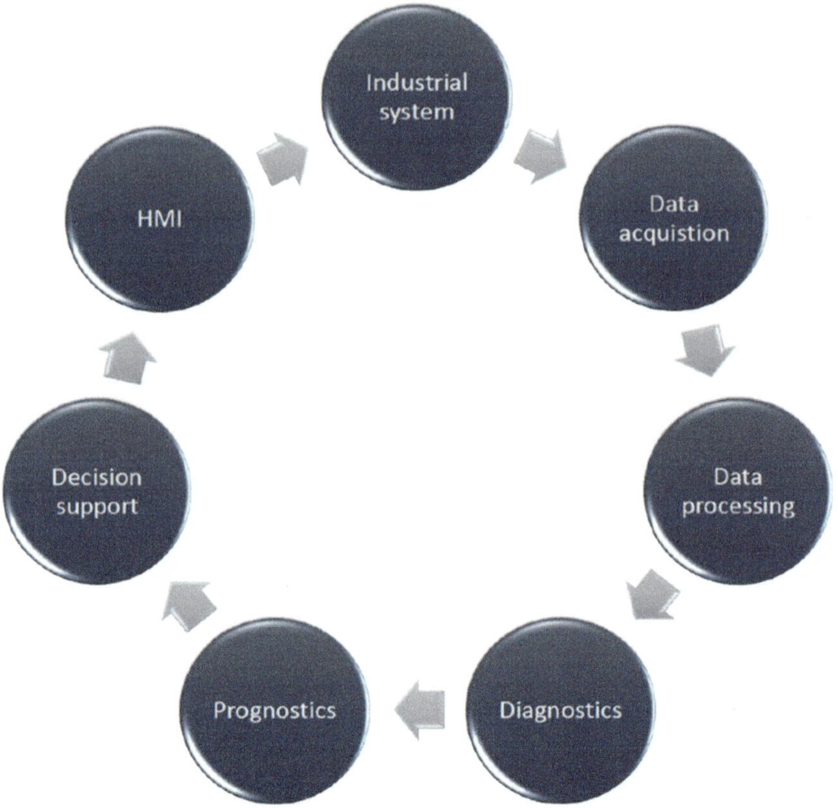

Fig. 8.1 Illustration of different steps in the PHM system [262]

PHM consists of seven steps initially defined as illustrated in Fig. 8.1. Each step is briefly discussed as follows:

(1) Data Acquisition and Processing: Data acquisition is the process of accumulating and storing raw sensor data related to the condition of the system. The data collected are usually categorized as CM data and event data. CM data are sensor measurements associated with system health, while event data are the knowledge obtained from an event (e.g., what kind of failure occurred, when and where the failure took place, who performed the maintenance procedure) [267]. Event data provides useful information on the performance of current features, as well as feedback on redesigning or improving features [268]. Thus, it is as important as CM data, although humans generally enter it manually, making it more fallible. An optimal maintenance system should automatically collect the event data.

Data processing includes data cleaning and data analysis. Cleaning isolates potential human and/or sensor faults and eliminates data that reflects these errors. The data can be a value type, a waveform type, or a multidimensional type [268]. Waveform and multidimensional data may contain noise. Therefore, cleaning also generally includes methods such as amplification, data compression, data validation, denoising, and filtering to enhance the signal-to-noise ratio. Data analysis extracts condition indicators that represent incipient and/or precursor failures or faults. The main purpose of these features is to maximize the accuracy of diagnostics and prognostics to decrease false alarms.

(2) Diagnostics and Prognostics: An effective PHM system includes diagnostics and prognostics approaches to provide ample and efficient decision support or automation. Diagnostics identify, localize and determine the severity of an evolving fault condition [269]. It involves fault detection, fault isolation, and fault identification. Fault detection, also called health/condition assessment, compares sensor data with expected operational performance, that is, expected values of system parameters such as pressure, temperature, and vibration, to identify irregular operating conditions. Fault isolation involves pinpointing the component or subcomponent that is degraded. Fault identification determines fault type and dimension according to classes associated with specific values of measured signals. Normally, this classification process uses a supervised classifier (e.g., machine learning algorithm) to classify various faults.

Prognostics predict the progression of faults and, hence, estimate the available time before a component or subcomponent loses its operational ability, that is, before a failure. Due to the large uncertainties involved, researchers have called the prognosis 'the Achilles heel' of PHM [270]. According to [271], the technical definition of prognostics is the estimation of time to failure (ETTF). However, in line with common usage in the literature, this paper uses the technical term remaining useful life (RUL). Any RUL estimate should include associated confidence intervals, which will indicate the window in which maintenance or repair must be performed [270]. Such intervals add assurance of continuous operation in spite of the inherent uncertainty associated with the degradation process, human errors, and flaws in both the diagnostics and prognostics approach applied in the PHM system [272]. Maintenance decisions based on prognostic information should be grounded in confidence intervals rather than a particular RUL value. The confidence intervals increase the reliability of the PHM system.

(3) Decision Support and HMI: The final object of a PHM system is to provide reliable decision support or automation to enable effective maintenance scheduling. Decision support should assist a decision maker (DM), while decision automation uses software to provide entirely autonomous decisions. However, according to [273], the output of today's industrial PHM systems usually constitutes decision support, as decision automation has not been integrated globally. Normally, inputs from human experts (application-dependent domain knowledge) and a DM who interprets the outputs compose the system [274]. Nevertheless, the human expert-generated input will fail when it encounters new conditions that the knowledge base does not define. Additionally, the most proficient DM has insufficient cognitive capacity to analyze and understand large amounts of information [275]. Hence, decision support in the age of big data is subjected to uncertainties and does not always ensure good quality decisions.

The advantages of a PHM system are highly connected to decision-making based on the accumulation and understanding of diagnostics and prognostic information. Making the best decisions based on large and complex amounts of information is difficult [275]. However, advanced and deep signal processing and machine learning techniques are evolving rapidly. These techniques provide automatic feature extraction and unsupervised learning procedures. Thus, such techniques minimize the input generated by human experts and have the potential to contribute to more intelligent PHM systems.

As [273] propose, the reliability of a fully autonomous (intelligent) PHM system must be greater than 99%. This level of reliability allows the PHM system to provide directions for maintenance procedures that are transferred directly from the system to maintenance personnel, without the involvement of a DM. Another important aspect of autonomous PHM systems is that they must prove reliability in order to use the 'black box' approach. Low-reliability PHM systems should allow the user access to source code to promote understanding and trust in the system [273].

The human-machine interface (HMI) is also an important perspective on understanding and trust, as screens and displays greatly affect how a DM or maintenance personnel understands the PHM system. A web-based PHM system, in which the user interacts using a thin-client web browser, has several advantages. According to [276], this system is powerful to retrieve, analyze, and visualize structured data from high-dimensional databases. It can provide access to unstructured data and promote communication and decision-making in distributed teams.

In the development of a PHM system, three distinct approaches are commonly employed: data-driven, physics-based, and hybrid. The hybrid approach integrates aspects of both data-driven and physics-based methodologies. However, this chapter focuses specifically on elaborating the data-driven approach. PHM systems have depended on traditional data-driven diagnostics and prognostics approaches [277, 278] and signal processing techniques [279]. With the development of the Internet of Things (IoT) and the rise of big data, traditional approaches face several challenges when processing the increasing volumes of data. Typically, they exploit human-engineered feature extraction methods, supervised machine learning algorithms, and shallow architectures. Therefore, traditional approaches are highly application dependent, require large quantities of labeled training data, and are simply not designed for large and complex datasets in real world applications [280].

However, during the past decade, with increased processing power and great progress in graphics processors, deep learning techniques have seen rapid developments. The areas of signal and information processing [281], speech recognition [282, 283], images [284], natural language processing [285], and visual tracking have seen significant improvements. Deep learning techniques consist of several layers of nonlinear processing stages. They utilize supervised or unsupervised learning strategies to automatically extract feature representations from raw input data. As a result, they can capture complicated, hierarchically statistical patterns in more complex, high-dimensional, and noisy real-world data [286]. For this reason, deep learning techniques are the most promising area of research to overcome the limitations of traditional diagnostic and prognostic approaches. There are several benefits and challenges to applying Prognostics and Health Management based on deep learning (PHMDL) in ships.

Benefits:

- Normally, critical systems on traditional ships are overengineered by built-in redundancy. In this way, traditional ships complete their operational tasks even if a serious functional failure occurs. This design philosophy is related to historical inaccessibility to shore. However, Inmarsat and Telenor have recently launched the data transfer satellites Inmarsat-5 and Thor 7, respectively, which will provide high-speed broadband connections to ships at sea [287]. This will enable new design philosophies, including online PHMDL systems, as alternatives to the legacy redundancy policy. Real-time diagnostics and prognostics of components and subcomponents in which online PHMDL systems are referred to an onboard system that links to shore will make it possible to contribute the most efficient operating conditions possible and enable future autoships without onboard maintenance personnel [288].
- The ultimate goal of a PHMDL system is to achieve zero-downtime performance. Real-time and reliable RUL estimations, with associated confidence intervals, of different components and subcomponents, will have an enormous impact on the maintenance procedure and safety concept on smart ships. When the RUL of a faulty component is estimated, the maintenance procedure can be scheduled to the next appropriate port of call, or if necessary, dispatching maintenance personnel before a failure occurs when the smart ship is still in operation. This will significantly increase the operational performance, and at the same time, drastically decrease unexpected system failures. In addition, reliable estimations will provide trust in safe behavior in offshore activities [289].
- According to [287], the insurance company Allianz reported in 2012 that between 75 and 96% of marine accidents are a result of human error. This is mainly a result of human exhaustion, but also because today's maritime activities require humans to manage planned operational activities and make complicated decisions based on the overall system conditions [289]. Autoships will reduce both the number of crew members and the influence of human DMs due to increased autonomous and intelligent operational planning and decision-making. In this way, autoships will have the potential to decrease human error and the risk of injury to crew members [288]. PHMDL systems have great potential to contribute to this human error reduction since these systems are less dependent on prior knowledge and human influence.

Challenges:

- Smart ship requires adaptation and integration within the functioning of a business of an organization, and hence, significant changes in the organizational culture [261]. The introduction of the smart ship also involves confidence and trust in 'black-box' systems, such as a PHMDL system. These systems are intelligent in that they transfer directions for future maintenance procedures directly from the

autoship to the maintenance team on shore. In order to act as a fully autonomous and intelligent system, the PHMDL system must adapt to the varying operational and environmental conditions that occur in the harsh maritime environment.
- A further concern is the continuous flow of data to shore. Autoships depend on heavily integrated and complex systems to deliver their main functions. As a result, the associated flow of sensor data becomes massive, high-dimensional, heterogeneous, and unstructured. The PHMDL system will have to provide automatic preprocessing and dimensionality reduction schemes. This massive flow of data also presents a cybersecurity challenge, as hackers would threaten safe maritime operations [287].
- A great challenge in the maritime industry is the lack of run-to-failure data on components and subcomponents. Traditional ships are often application designed and unique, or batch produced in two to ten vessel series [261]. These short series create a slow accumulation of relevant failure data compared to, for instance, the aviation industry that produces hundreds of the same aircraft in series [4]. In addition, traditional ships are typically equipped with components from several different manufacturers. The resulting diversity of uncoordinated monitoring systems increases the complexity of the failure data. With respect to the introduction of smart ships and PHMDL systems, it would be advantageous to build extensive databases regarding run-to-failure data of critical and relevant components and subcomponents. This could be realized if stakeholders agreed to cooperate to share data.

In this chapter, four primary actions are undertaken within a comprehensive deep learning-based PHM system: anomaly detection, fault diagnostics, fault prognostics, and, ultimately, decision support or automation.

8.2 Anomaly Detection

Anomaly detection focuses on identifying instances or events that deviate significantly from the expected or normal behavior within a given context, irrespective of the underlying cause. This technique has been widely used for fault detection of various machines because, in general, faults generate signals with characteristic patterns that differ from normal conditions and indicate changes in machine behavior. Faults can occur due to component failures, environmental changes, parameter drifts, or operational errors. In this context, fault/anomaly detection aims to identify whether the state of a machine deviates from its normal state. In the simplest case, sensor data is compared with expected operating performance (i.e., expected values of system parameters such as pressure, temperature, and vibration) to identify irregular operating conditions. However, a machine may involve a large number of sensors, and some of them are correlated, making it difficult to set thresholds for each sensor measurement or define the state of the machine. From a machine learning perspective, fault/anomaly detection can be defined as building a **binary classifier** to classify whether a machine is in a normal state or an abnormal state.

8.2.1 Type of Anomalies

To build an anomaly detection model, an important aspect is to understand the nature of the desired anomalies. Given time series data, anomalies can be divided into the following three categories (shown in Fig. 8.2):

- *Point anomalies.* This is the simplest type of anomaly. An individual data point in the time series can be termed a point anomaly if it is anomalous with respect to the rest of the data points.
- *Contextual anomalies.* If a data point is anomalous in a specific context of the time series, but not otherwise, then it is termed a contextual anomaly. As demonstrated in Fig. 8.2, point a is characterized as a contextual anomaly as it happened before but now occurs in the wrong context. The difference to point anomaly is that point anomaly is context-independent. The notion of a context is induced by the structure of the dataset. For example, in a manufacturing process, time-series data may be strictly time-dependent, since each production step is fixed and thus time is the context. For marine diesel engines, boost pressure is positively related to engine speed, a contextual anomaly for boost pressure is that boost pressure decreases when engine speed increases, here engine speed is the context.
- *Collective anomalies.* If a collection of related data points is anomalous with respect to the rest of data, it is termed a collective anomaly. Individual data points in a collective anomaly may not be anomalous by themselves, but they are anomalous when they occur together as a collection.

It should be noted that these three types of anomalies can overlap. A point anomaly can be a contextual anomaly and can be included in a collective anomaly. A collective anomaly can be a collection of point and contextual anomalies. In the fault detection context, the anomaly is usually a combination of collective and contextual anomaly. This is because faults do not appear and disappear suddenly, and most faults need to be identified from multiple sensor measurements rather than univariate time series. Even though early time series anomaly detection focuses specifically on one type of anomalies [290, 291], modern deep neural network-based models that take the time series data as input instead of a data point in time series usually do not assume the type of anomalies and can be used easily in the fault detection context.

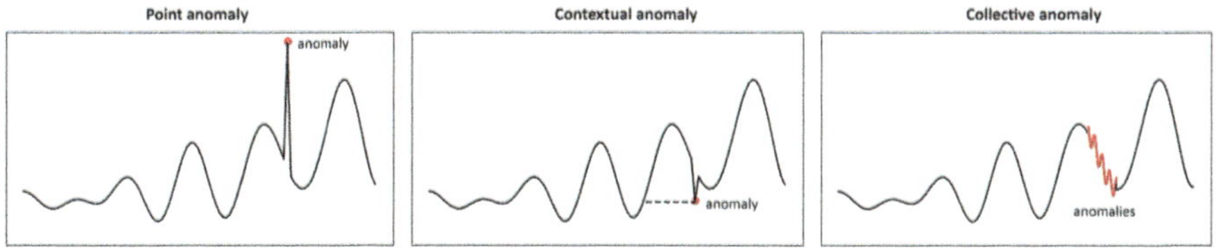

Fig. 8.2 Schematic illustration of different types of anomaly

8.2.2 Categories of Anomaly Detection Methods

Building an anomaly detection model based on machine learning requires data and labels. The labels associated with the data points denote whether the data points are normal or anomalous. Labeling the data is often done manually by human experts. There is no doubt that obtaining accurately labeled data representing all possible behaviors requires substantial effort. In addition, getting a labeled dataset of anomalous data that covers all possible types of anomalous behavior is much more difficult than getting a labeled dataset of normal data. This could be due to the fact that anomalous behavior is rare. Based on what kind of label is available, the anomaly detection methods can be divided into the following three categories (as shown in Fig. 8.3. The orange and blue points denote normal and anomalous data, respectively. The dashed line denotes the possible decision boundary learned by the model):

- *Supervised anomaly detection.* This method assumes that the training dataset has both normal and anomalous data. A typical approach is to train a binary classifier to classify normal and anomalous samples. A major concern for supervised anomaly detection is the data imbalance problem since abnormal data is much less than normal data. For neural network-based classifiers, resampling techniques or cost-sensitive learning can be used to solve the problem.
- *Semi-supervised anomaly detection.* This method assumes that the training dataset has only the normal data, that is, no labeled anomalous data is available. The approach usually assumes that the anomalous data is far away from the normal data. It is more widely used than supervised anomaly detection, but it can be problematic when anomalous data is similar to normal data.
- *Unsupervised anomaly detection.* This method assumes that the training dataset has no labeled data, that is, the dataset contains both normal and abnormal data, but the labels are not available. The approach usually assumes that normal data are far more frequent than anomalous data. Many semi-supervised approaches can be adapted to unsupervised anomaly detection since the model is usually robust to a few anomalous data. If anomalous data is dense, it could be characterized as normal.

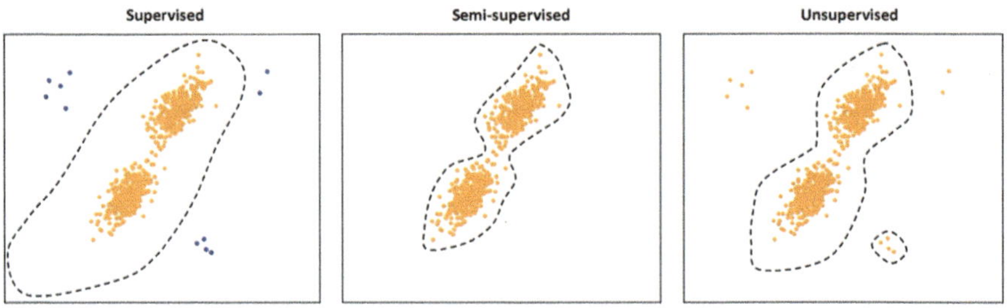

Fig. 8.3 Schematic illustration of different categories of anomaly detection methods. The orange and blue points denote normal and anomalous data, respectively

In the context of fault detection, semi-supervised methods are most applicable because normal data are usually readily available. This is the same for maritime components such as the diesel engine onboard. Recording anomalous data is costly or even dangerous in comparison to normal data. Semi-supervised anomaly detection models include a one-class support vector machine, isolation forest, principle component analysis, etc. However, these methods focus on structured data and when applied to time series data, extensive feature engineering must be performed. In this chapter, we only focus on deep neural network-based anomaly detection as it can easily handle time series data.

8.2.3 Neural Network-Based Semi-supervised Anomaly Detection

Deep neural network architectures such as 1D Convolutional Neural Networks, Recurrent Neural Networks, and Transformers have demonstrated their superior performance in modeling time series data, ranging from univariate to multivariate data. Measured data from diesel engines or other machinery and electronic components on a ship, usually consists of several different sensors such as temperature, vibration, etc. These measurements are therefore multivariate time series data which can be handled by the deep neural network above. Given a dataset contains only normal multivariate time series data, the principle of neural network-based anomaly detection can be divided into two categories (as shown in Fig. 8.4):

- *Forecasting-based model*. The forecasting-based models train a neural network to forecast future time series based on past time series. The idea is that a model can only predict normal future time series when trained on a normal dataset, while the model cannot predict anomalous patterns. The model can detect anomalies based on prediction errors.
- *Reconstruction-based model*. The reconstruction-based model leans a neural network model to capture the latent representations (low-dimensional representations) of the given time series and then reconstruct the time series based on the latent representation. The model assumes that anomalies lose information when they are mapped to a lower dimensional space and thereby cannot be effectively reconstructed. Therefore, the reconstruction errors can be used to detect anomalies.

It should be noted that prediction and reconstruction errors are general terms here, which can be mean square error or even reconstruction probability [292] when a variational autoencoder is used as the reconstruction model. The forecasting-based and construction-based models can also be used complementarily for anomaly detection [293].

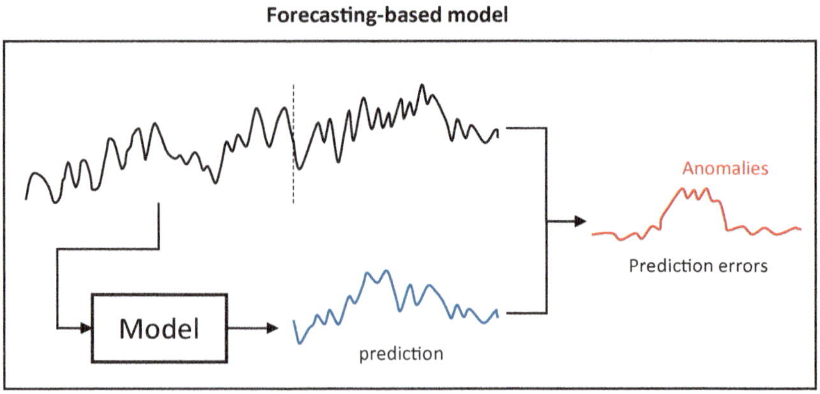

(a) Forecasting-based model.

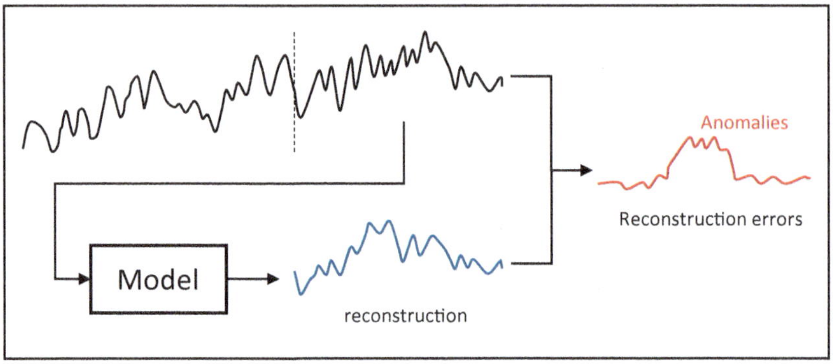

(b) Reconstruction-based model.

Fig. 8.4 Schematic illustration of different neural network-based semi-supervised anomaly detection models

8.2.4 Case: Fault Detection for Marine Diesel Engine

In this section, a marine diesel engine is used as an example to demonstrate how to use a reconstruction-based model to perform anomaly/fault detection. The marine diesel engine is one of the most critical components onboard ships since it has an important role in both propulsion and power generation. It is subjected to rapid variations in operational loads, depending on both the task of operation and environmental conditions. In such complexity, some of the anomalies cannot be presented directly for a single sensor measurement since the sensor measurements are highly connected to the operational loads. In addition, it is also difficult to manually identify the anomalies since there are multiple sensors involved and the change of sensor measurement might not be obvious.

8.2.4.1 Marine Diesel Engine and Data Collection

The diesel engine used as an example is operated on the Norwegian University of Science and Technology's research vessel (RV) Gunnerus. RV Gunnerus is equipped with the latest technology for a variety of research activities within biology, technology, geology, archaeology, oceanography, and fisheries research. The diesel-electric system of RV Gunnerus is used to generate electric power which is supplied to the power grid for operating the vessel. There are three engines on board, usually two running and one as a backup. One of the running engines as shown in Fig. 8.5 is used to demonstrate anomaly detection.

The ship's diesel engine works normally most of the time. This is due to the current preventive maintenance strategy, which is relatively conservative but not prone to faults. It is therefore easy to obtain engine normal operation data since it is normal operation data as long as the engine is running. We collected the data from the entire month of November 2019. During these periods, the vessel has been sent out for several purposes such as sea trials, maneuvering courses, etc. No specific fault for the engine was found in this period. The time interval when the vessel is in operation is filtered out. In total, we got 10 d that the vessel is in operation and approximately an average of 6 h for each day. Table 4.1 lists the sensor measurement related to the diesel engine from the logging system. The sensor data was collected at a sampling rate of 1 Hz.

To validate the model, anomalous data is essential. On 21st, November 2019, we went on board to introduce a fault in this diesel engine. The air filter clogging fault

Fig. 8.5 Diesel engine operated in the NTNU's research vessel. The air filter is manually clogged for a period of time

was simulated using a cloth winding tape, as shown in Fig. 8.5. The left subgraph in Fig. 8.5 shows the diesel engine onboard and the right subgraph presents that the air filter is clogged with the tape. The outer surface of the air filter of the diesel engine is wrapped with a cloth winding tape. In this way, the heat dissipation and exhaust capacity of the air filter are reduced. Note that the fault introduced can be categorized as an abrupt fault.

The 10 d of collected data are divided into the following three parts: (1) a test set containing 2 d of data: the day when the fault was introduced and a random normal operation day; (2) a validation set containing data for 1 of the remaining 8 d; (3) a training set containing the data for the remaining 7 d. The training set and validation set are used in the training phase of the model. The test set is used to show the efficacy of the model. During these 10 d, the vessel is operated around the fjord of Trondheim. Since the training, validation, and test data come from different operation days of the ship, resulting in different operating conditions, data leakage is unlikely to happen. Furthermore, the test result can reveal that the proposed model can detect the fault regardless of the operation conditions.

8.2.4.2 Data Normalization

Each sensor measurement in the training data sets is scaled with standard (z-score) normalization:

$$\overline{x}_n = \frac{x_n - \mu}{\sigma} \tag{8.1}$$

where μ and σ is the mean and standard deviation, respectively. n refers to the sensor index. The normalization statistics obtained from the training set are then applied to the validation set and test set, respectively.

Methodology: Long-Short Term Memory Based Variational Autoencoder

We introduce a long short-term memory-based variational autoencoder (LSTM-VAE) for anomaly detection in our case. LSTM-VAE is a combination of VAE and LSTM. In particular, LSTMs are used to replace the encoder $q_\phi(z|x)$ and decoder $p_\theta(x|z)$ modeled by feed-forward networks in VAE, thereby introducing temporal dependencies.

Figure 8.6 shows the LSTM-VAE structure that is unrolled in time. Note that FC is a fully connected layer. The FC in the encoder has Relu activation while the FC in the decoder has identity activation. LSTM uses tanh activation. η is the fault detection threshold. Given a multivariate input x_t at time t, the encoder LSTM output the hidden state h_t utilizing x_t, h_{t-1}, c_{t-1}. Then h_t is fed into two linear modules to estimate the mean μ_t and log-variance $\log \sigma_z$ of the posterior $p(z_t|x_t)$. A random vector z_t from $p(z_t|x_t)$ feeds into the decoder LSTM and then a final linear module outputs

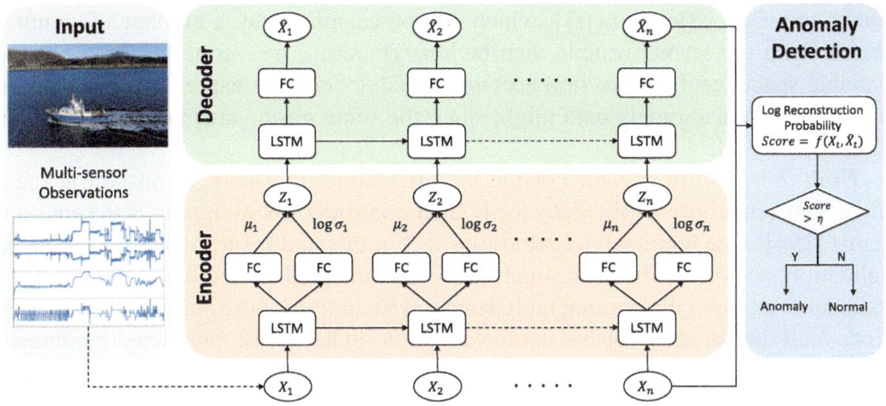

Fig. 8.6 Illustration of the LSTM-VAE anomaly detector unrolled in time

the reconstructed input $\hat{x}_t$. At each time step, the model can be viewed as a VAE. The LSTM connects the VAE at each time step and thus the temporal dependencies are considered.

The parameters ϕ for the encoder and θ for the decoder can be obtained by minimizing the loss function as follows:

$$Loss = \sum_{t=1}^{T} [D_{KL}\big(q_\phi(z_t|x_t)||p_\theta(z_t)\big) + MSE(x_t, \hat{x}_t)] \tag{8.2}$$

where MSE denotes mean square error, T is the length of the sequences. A standard normal distribution $\mathcal{N}(0, 1)$ is used as the prior $p_\theta(z_t)$ of the latent space. A multivariate Gaussian distribution can be assumed for continuous data and therefore maximizing the log-likelihood equals minimizing the MSE in Eq. (8.2).

Both the encoder LSTM and decoder LSTM consist of two-layer LSTM with hidden units number $\{12, 8\}$. The size of the latent space is selected as 4. The proposed LSTM-VAE is trained using the back-propagation algorithm. Since one day of data usually contains over 25,000 time steps, it is difficult to train LSTM with such a time length. The training data in each day was cut into segments with 1,000 data-point using stride 10 and then the batch learning can be applied. The mini-batch size is set to be 512 and ADAM is used as the optimizer with a learning rate of 1×10^{-3}. The l_2 regularization term with coefficient 1×10^{-3} was used. The training process stops if the loss for the validation set with no decrease for 100 epochs. The hyperparameters are selected by trial and error and these values provide satisfactory results for the training of the proposed model.

Anomaly score. In general, mean square error is usually used as the anomaly score. Since VAE is stochastic in nature, the variability of the latent space can be taken into account. We use the reconstruction probability as the anomaly score for the proposed LSTM-VAE. The reconstruction probability is the Monte Carlo estimate of the log-likelihood $E_{q_\phi(z|x)}[\log p_\theta(x|z)]$, which can be calculated by a number of samples drawn from the latent variable distribution. Therefore the variability of the latent variable space can be taken into account, which extends its expressive power since normal data and anomaly data might share the same mean value but have different variability [292].

Fault detection threshold. For the fault detection threshold, a constant is specified. The sensitivity of the detector is then controlled by assigning different constants [294]. One approach to determine such a threshold is to maintain a separate validation set with fault and normal data. The threshold can be tuned based on the validation set to get the optimal fault detection accuracy or false alarm rate. However, since fault data is not available in the validation set here, The fault detection threshold is determined as 99.7% of the anomaly score in the validation set that contains only the normal data. The assumption is that the anomaly score should have a similar range as the validation set when the component is normal. We empirically found that this threshold provides satisfactory results in our case.

Experimental Setup

Evaluation metrics. Two performance evaluation metrics, *time to detect* and *detection stability factor* [295], are used in this paper to evaluate the performance of the method. *Time to detect* (TTD) is defined as the period of time from the beginning of a fault injection to the moment the first detection signal occurs. *Detection stability factor* (DSF) is the level of stability of the detection signal measured as a percentage of the sum of the duration of fault detection signals to the total time elapsed after fault injection. Figure 8.7 presents a schematic illustration of the fault detection process, where t_{f1} is the moment when the fault is introduced while t_{f2} is the time that the fault is ended. These two evaluation metrics can be calculated through Eq. (8.3).

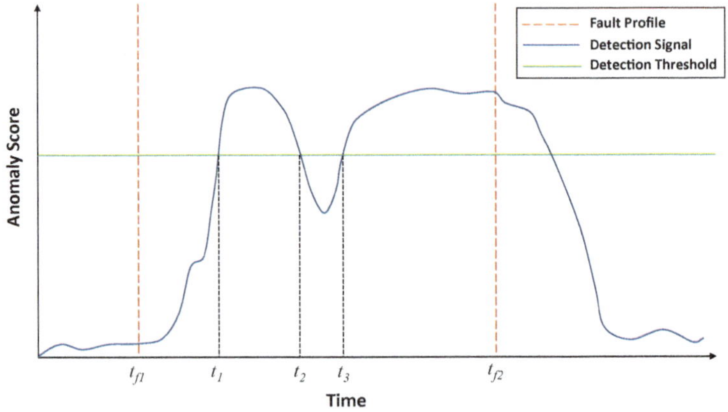

Fig. 8.7 Illustration of the evaluation metrics for fault detection of marine diesel engine

The *time to detect* attempts to measure how quickly the fault detection algorithms respond to faults while the *detection stability factor* measures the stability of fault detection.

$$\begin{aligned} TTD &= t_1 - t_{f1} \\ DSF &= \frac{(t_2 - t_1) + (t_{f2} - t_3)}{t_{f2} - t_{f1}} \end{aligned} \tag{8.3}$$

Baseline methods. To evaluate the performance of the proposed method, we implemented 6 baseline methods:

- iForest: An isolation forest-based detector with standard normalization.
- iForest (MRN): An isolation forest-based detector with multi-regime normalization (MRN).
- AE: An autoencoder-based detector with standard normalization.
- AE (MRN): An autoencoder-based detector with multi-regime normalization.
- VAE: A variational autoencoder-based detector with standard normalization.
- VAE (MRN): A variational autoencoder-based detector with multi-regime normalization.

For a fair comparison, the hidden unit size and the latent space size for both AE and VAE are set the same as LSTM-VAE. Multi-regime normalization [296] refers to normalizing the data based on its operation conditions. Since no prior knowledge about the number of operating conditions or the way to divide the data into different operation conditions is available, we perform multi-regime normalization through three steps: (1) Calculate the coefficient of variation (CV) for different sensors and select four sensors with the highest CV; (2) Perform K-Means clustering based on these four sensors; (3) Normalize all the sensors measurements based on the cluster it belongs. The first step is to select the relevant sensor measurements for the operation condition. The operation condition is approximated by the cluster.

Experimental Results

Training, validation, testing loss. Table 8.1 shows the loss for different methods. The testing loss is calculated from the day when no fault is introduced to the engine. The loss of isolation forest-based anomaly detector is the anomaly score defined in [297]. The loss of AE is the reconstruction error while the loss of VAE and LSTM-VAE is the KL-divergence plus the reconstruction error. It is shown that the training, validation, and testing loss are in the similar range for these methods. This indicates that the models are well trained and the training, validation, and testing data are from the same distribution.

Qualitative results. Figure 8.8 shows the qualitative comparison of our proposed LSTM-VAE with the baseline methods. The left five subgraphs are the anomaly scores from the day where a fault is introduced. The right five subgraphs show the

Table 8.1 Training, validation, testing loss of different methods

Method	Training	Validation	Testing
iForest	0.437	0.454	0.471
iForest (MRN)	0.402	0.410	0.450
AE	0.007	0.014	0.022
AE (MRN)	0.114	0.134	0.329
VAE	0.358	0.436	0.504
VAE (MRN)	0.772	0.820	1.195
LSTM-VAE	0.486	0.496	0.503

anomaly score from the normal operation test day. The red background on the graphs represents the ground truth of the fault. The fault detection algorithms are run on the test set to produce the anomaly score. The left subgraphs are from the day when a fault is introduced while the right subgraphs are from the normal operation test day. The period where the air filter clogging fault is introduced is marked with a red background. The anomaly score is filtered by a median filter with kernel size 79 to remove the random spike.

From the left first three subgraphs, the fault can be only detected when the multi-regime normalization is used. When standard normalization is used, there is no distinguishable increase in anomaly score at the fault time step for AE and VAE. Even though the anomaly score is noticeable at the fault time step for iForest, the score in normal operation is relatively high. The right four subgraphs show the anomaly score for one normal operation day. The scores are therefore relatively low.

The left fourth and fifth subgraphs in Fig. 8.8 show that the LSTM-VAE provides a similar result to iForest, AE, and VAE when the multi-regime normalization is used. A clear increase in reconstruction error can be found at the fault time step. The LSTM-VAE applies directly to the standard normalized data, which makes the method easy to scale to a complex system. Generally, lots of sensors are equipped in a maritime system, and complex operation conditions are involved. Performing multi-regime normalization is unrealistic in most scenarios. Even for this diesel engine, we spent lots of effort deciding the relevant sensor and the number of clusters. LSTM-VAE naturally includes the temporal dependencies and there is potential to easily scale the model in maritime systems.

From the left fourth and fifth subgraphs in Fig. 8.8, it is shown that the reconstruction probability provides a more noticeable change than the reconstruction error. It indicates that it is beneficial to include variety in the latent space. The reconstruction probability is expected to be more expressive than the reconstruction error.

Quantitative results. Table 8.2 summarizes the performance of different methods in terms of time to detect (TTD) and detection stability factor (DSF). Different kernel size ω in the median filter is used to smooth the anomaly score. Only the results of the Iforest, AE, and VAE with the multi-regime normalization are shown since these models with standard normalization fail to detect the fault. It is shown that the AE, VAE, and LSTM-VAE perform better than the iForest in our case. The AE, VAE, and LSTM-VAE show a similar performance with TTD around 80 s and DSF around

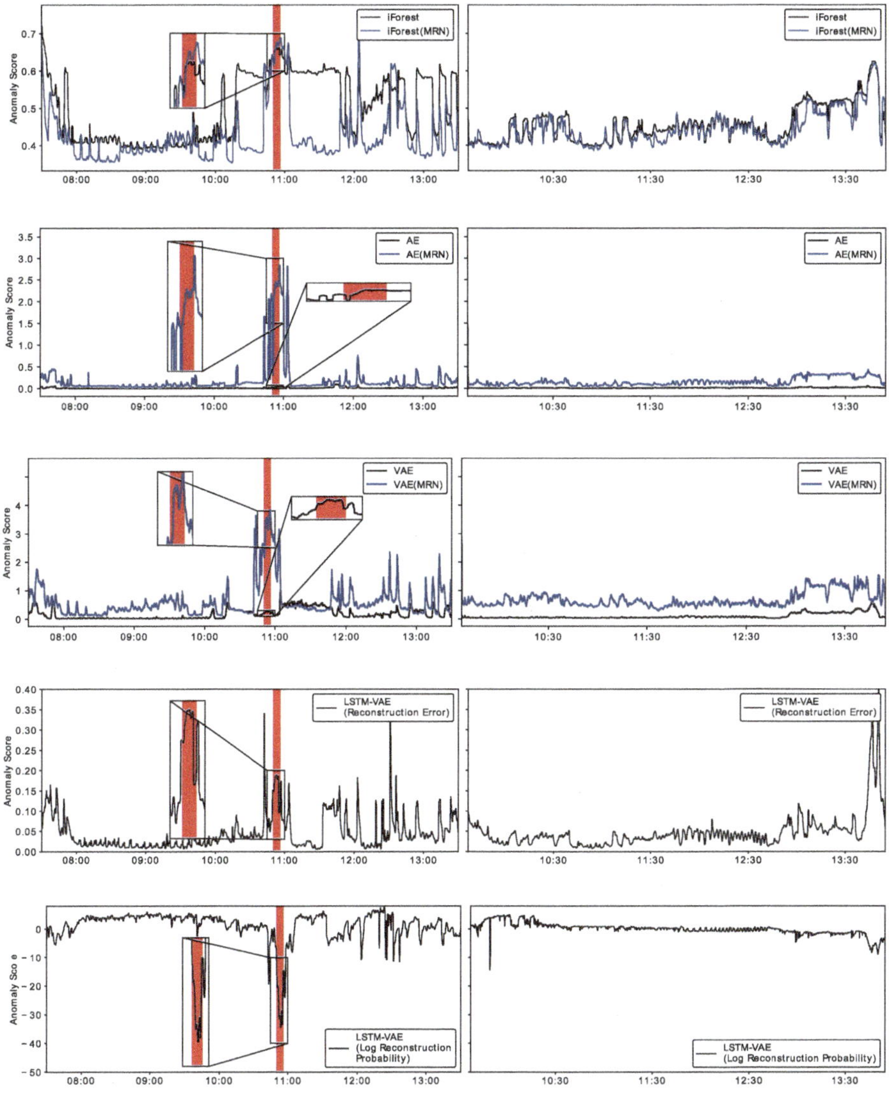

Fig. 8.8 Visualization of the anomaly scores over time in the test set

0.78. For the LSTM-VAE, it is shown that using log reconstruction probability as anomaly score provides lower TTD as well as higher DSF than using reconstruction error. With log reconstruction probability, the LSTM-VAE can archive TTD as 60 s and DSF as 0.791.

It should be noted that the current work only considers two temporal metrics to evaluate the proposed method, which is used to demonstrate that the model can detect anomalies. More data including different failure types under different operating

Table 8.2 Comparison of different methods

Method	ω	TTD ↓	DSF ↑
iForest (MRN)	39	163	0.553
	59	175	0.564
	79	178	0.565
	99	182	0.565
AE (MRN)	39	77	0.755
	59	78	0.781
	79	78	0.781
	99	79	0.781
VAE (MRN)	39	76	0.757
	59	80	0.749
	79	83	0.783
	99	84	0.785
LSTM-VAE (Reconstruction Error)	39	75	0.782
	59	77	0.782
	79	80	0.784
	99	83	0.782
LSTM-VAE (Log Reconstruction Probability)	39	**60**	0.774
	59	65	0.786
	79	66	**0.791**
	99	72	**0.791**

conditions should be included, and the precision, recall, and F1 score used in standard binary classification tasks should be used to provide a comprehensive evaluation.

8.3 Fault Diagnostics

Fault diagnostics involves the process of understanding the root causes of these faults and identifying the degree of faults, which is usually called fault isolation and fault identification, respectively. By accurately diagnosing faults, engineers, technicians, or automated systems can take appropriate corrective actions to rectify the issues and restore system functionality.

From the perspective of the data-driven method, fault isolation is a multi-class classification problem, and fault identification could be a classification or regression problem. The fault isolation model takes the sensor measurements as inputs and outputs the cause of the fault, e.g., identifying whether the fault is from the air filter, turbocharger, or fan in the cooling system for a marine diesel engine. For systems such as ships, fault isolation might aim to isolate which thruster failed [298]. As for fault identification, the degree of fault can be defined as categorical or continuous values. For a popular bearing fault dataset (Case Western Reserve University (CWRU) dataset), Single-point faults are introduced to the bearings under test using electro-discharge machining with fault diameters of 7, 14, 21, 28, and 40 mils, at the inner raceway, the rolling element, and the outer raceway. Thus fault identification is a classification problem. For batteries, fault identification could refer to estimating the state of health. In such a context, fault identification is a regression problem and it is closely related to health state estimation.

The fault diagnosis problem is usually viewed as a classification or regression problem using various deep-learning models in the literature. One of the key issues in using deep learning models for fault diagnosis is the lack of fault data. Another issue is that changes in operating conditions may cause changes in the distribution of the data. Both issues are under active investigation. Because we do not have sufficient data on marine engine failures, fault diagnosis is not discussed further in this chapter.

8.4 Fault Prognostics

Fault prognostics go beyond traditional fault detection and diagnosis by providing an early warning of potential failures, enabling proactive maintenance and minimizing downtime. Prognostics predict the progression of faults, and hence, estimate the available time that remains before a component, system, or asset reaches a predefined threshold or fails to meet its intended performance requirements. It is also called the remaining useful life (RUL) prediction. Because of the large uncertainties involved, researchers have called prognostics "the Achilles' heel" of PHM [262]. The formal definition of remaining useful life can be expressed as follows:

$$RUL(t) = T - t \tag{8.4}$$

where $RUL(t)$ is the remaining useful life at current time t. T represents the time of failure. It's important to note that estimating the Remaining Useful Life is often a dynamic and iterative process that involves updating the RUL estimation based on real-time data, system health monitoring, and predictive models. As new data becomes available, the RUL estimation can be refined and adjusted, providing a more accurate assessment of the remaining time until failure.

8.4.1 Principles of Data-Driven Fault Prognostics

The data-driven fault prognostics have received increasing attention in recent years. However, building such a data-driven model requires run-to-failure data. The asset has to run until the failure occurs so that the degradation process is recorded. This can be very expensive and time-consuming, so currently the run-to-failure data is rather limited. Given a run-to-failure dataset, an RUL prediction model is usually built under the following two principles (as shown in Fig. 8.9):

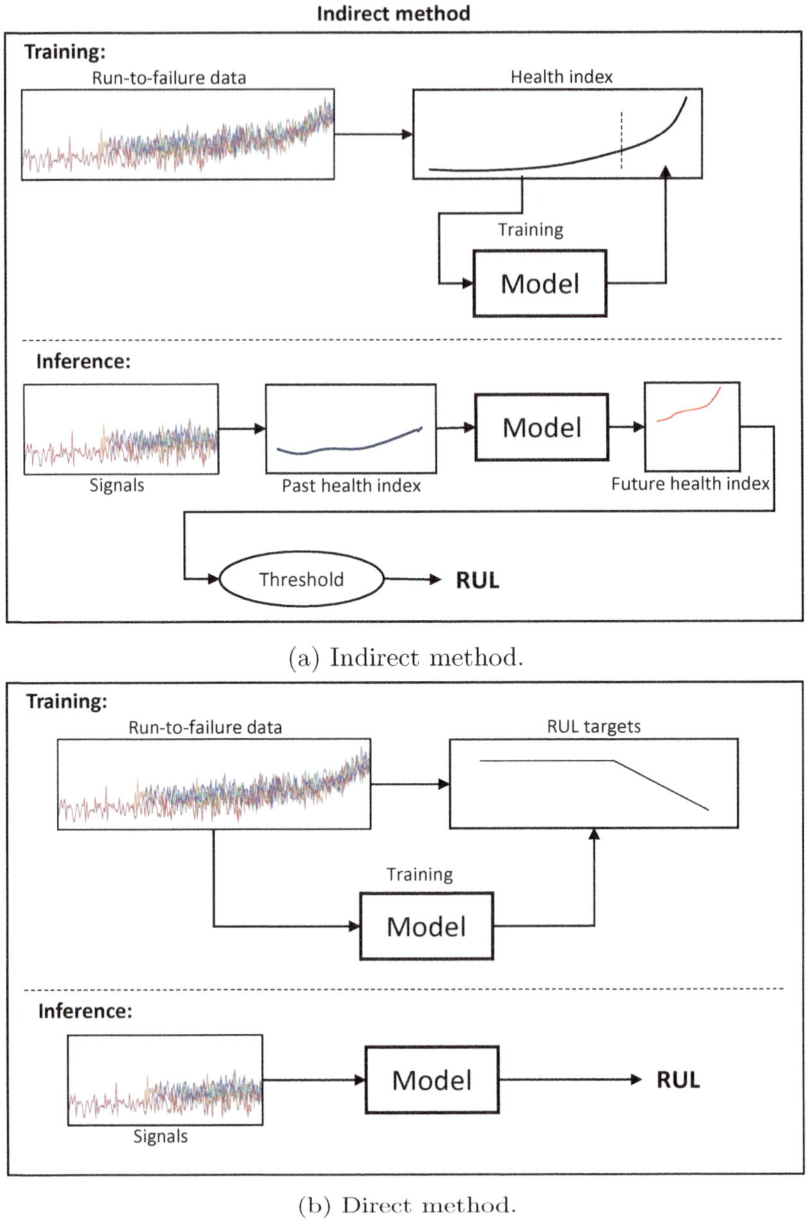

Fig. 8.9 Schematic illustration of different RUL prediction models

- *Indirect method.* The model based on this principle provides the RUL indirectly. The method first constructs the "health index" from the run-to-failure signals. The "health index" can be explicitly defined as the state-of-health (SOH) of a battery, or it can be implicitly chosen based on the physics-of-failure analysis, such as the radio-frequency impedance being used to indicate the health conditions of interconnects [299]. Furthermore, the "health index" can also be the result of fault identification in the previous section. The "health index" can then be constructed from all samples run to failure in the dataset, and the propagation of this "health index" can be learned using a data-driven model, usually formulated as a time series forecasting problem, e.g. with the past "health index" to predict the future "health index". Thresholds for "health indices" indicative of end-of-life may also be determined. In this way, the RUL is calculated by propagating the "health index" to its threshold.
- *Direct method.* The model based on this principle provides the RUL directly. This method assigns the RUL targets to all the run-to-failure data. Data-driven models are used to learn to directly predict RUL. Thus besides the model, a key to this approach is how to construct the RUL targets. The commonly used RUL targets are linear and piece-wise linear targets. One can also use a similar idea to the "health index" above to fit polynomial targets from descriptive sensors. Figure 8.10 shows a comparison of these three RUL targets.

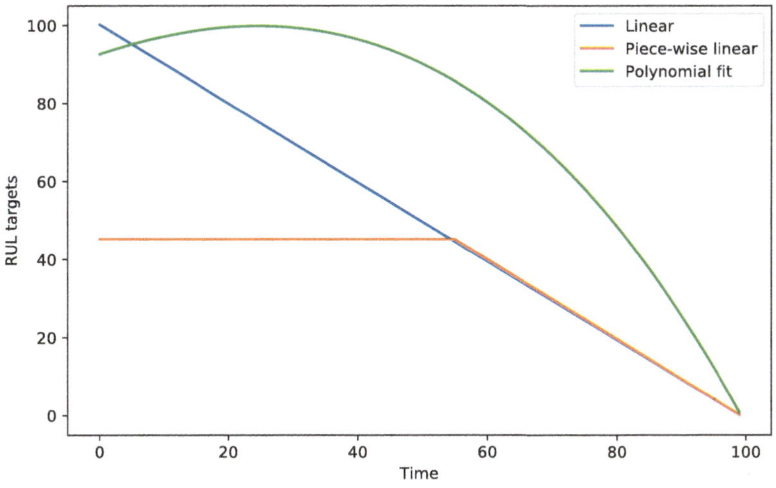

Fig. 8.10 Comparison of three different RUL targets

8.4.2 Evaluation of Fault Prognostics Models

To evaluate the performance of fault prognostics models, a separate test dataset that contains the ground truth RUL (RUL_{target}) is needed. Two evaluation metrics are usually used as performance indicators: the root mean square error (RMSE) and the maintenance scoring function (SCORE).

The RMSE is defined as follows:

$$RMSE = \sqrt{\frac{1}{n}\sum_{i=1}^{n} d_i^2} \tag{8.5}$$

where n is the total number of constructed RUL targets and $d_i = RUL_{predicted,i} - RUL_{target,i}$.

The maintenance scoring function (SCORE), provided in [300], is defined as follows:

$$SCORE = \begin{cases} \sum_{i=1}^{n} e^{-(\frac{d_i}{13})} - 1, & \text{for } d_j < 0 \\ \sum_{i=1}^{n} e^{(\frac{d_i}{10})} - 1, & \text{for } d_j \geq 0 \end{cases} \tag{8.6}$$

In SCORE, the punishment for late RUL predictions, when $d_j > 0$, is greater than early RUL predictions when $d_j < 0$. Late RUL predictions are vulnerable to failure since maintenance operations would be scheduled too late. Early RUL predictions, however, pose less risk of engine failure as maintenance operations would be scheduled too early. Additionally, SCORE is a suitable metric for real-time predictions since it only involves the latest input data and corresponding RUL prediction.

The main objective for both performance indicators is to reach the lowest value possible, namely when $d_i = 0$ and $d_j = 0$. The RMSE and the SCORE are illustrated in Fig. 8.11.

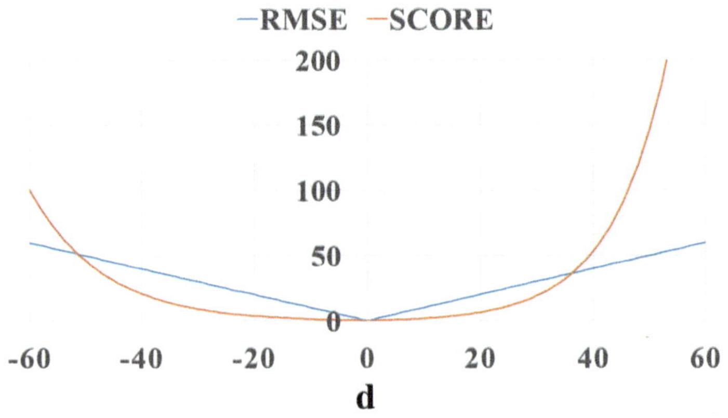

Fig. 8.11 RMSE versus SCORE

8.4.3 Case: Fault Prognostics of Marine Diesel Engine

In this section, a marine diesel engine is used as an example to demonstrate how to use the direct method to perform fault diagnostics.

Run-to-Failure Data Collection

A hybrid power lab, which is designed to research ship autonomy, is used to collect the run-to-failure data sets for fault prognostics. The lab is established by the Department of Ocean Operations and Civil Engineering at the Norwegian University of Science and Technology in Aalesund. The lab includes a marine diesel engine, a marine battery system, and a marine automation system to control the process. To simulate load alterations in the system, the produced power is supplied back to the power grid. The diesel engine is shown in Fig. 8.12.

Two engine load profiles have been used during the data collection process. The profiles aim to replicate two different environmental conditions the autonomous ferry may encounter during a ferry crossing on the west coast of Norway. First, the engine runs in normal condition. Then, two typical and independent fault-types associated with the marine diesel engine were provoked gradually in both profiles to obtain degradation data. The first fault-type is clogging of the air filter, while the second fault-type is a malfunction of the turbocharger. The data collection process was terminated when the engine reached operational failure. That is, a time after the fault-types were 100% provoked and the engine loses its operational ability.

Profile 2 is subjected to different engine loads, and hence, it will be used as the test set. However, the degradation in the test set should end sometime before failure in order to verify that the model is able to predict the RUL. Accordingly, a random interval of time steps is removed in both the air filter fault and the turbo fault in profile 2. Table 8.3 summarizes the data sets used. All data sets include 47 input features, e.g., engine load, engine speed, flow, pressure, and temperature measurements.

Methodology

RUL targets construction. The piece-wise linear (PwL) degradation model is widely used to label the RUL targets. In the original PwL model, all engines utilize the same initial RUL values. Since the engines might follow different degradation patterns, an improved PwL model as presented in [301] is used.

Figure 8.13 depicts the improved PwL model. An anomaly detector is used to detect the time step where the fault is initially provoked. Then the RUL targets are constructed based on the PwL model: constant RUL targets before the fault time step and linearly degraded RUL targets after the fault time step. Consequently, the last RUL target = 0.

Fig. 8.12 The marine diesel engine in the hybrid power lab

Table 8.3 The four original run-to-failure data sets collected from the marine diesel engine

Data set	Profile	Usage	d_{ft}	Last RUL target	Time steps
Air filter	1	Train/val	1,660	0	2,346
Turbo	1	Train/val	1,347	0	2,346
Air filter	2	Test	1,483	106	2,240
Turbo	2	Test	1,347	490	1,856

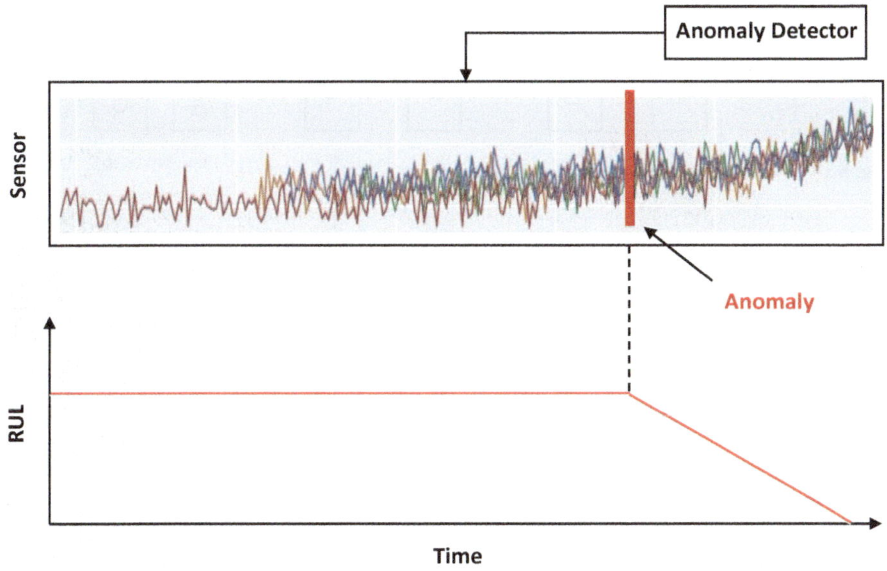

Fig. 8.13 Illustration of the improved piece-wise linear degradation model

Data augmentation. First, random white Gaussian noise is added to each sensor measurement in the original training set with a random signal-to-noise ratio between 70 and 90%. Next, following [302], a random interval of time steps is removed after the fault time step in the RUL targets construction to also include some time series that will end some time prior to failure. The augmented data will be used to train the model.

Network Architecture. The LSTMs and feed-forward neural networks (FNNs) are used as the main building blocks for RUL prediction. The LSTM layers are used to learn temporal and long-term dependencies within the features of degradation data. The FNN layers are then used to map all extracted features before a dropout layer is used to reduce overfitting. Dropout [303] randomly drops units of the dropout layer during training. This forces the model to learn to construct generalized representations of the input data. The last layer consists of a fully connected output layer with one unit. This layer handles error calculations and performs RUL predictions. The mean squared error is utilized as the loss function.

The used network architecture can be seen in Fig. 8.14. Two layers of LSTM and two layers of FNN are used, with skip-connection on the last layer of LSTM and FNN, respectively. The skip connections in the model use element-wise addition to combine the activations of two layers.

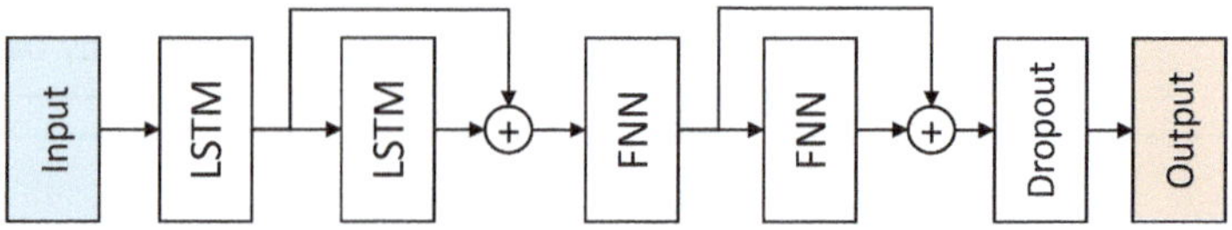

Fig. 8.14 Network architecture for RUL predictions

Table 8.4 Comparison with different approaches for engine RUL predictions

Approach	RMSE	SCORE
CNN+FNN [304]	96.26	40.32
LSTM+FNN [305]	285.09	67.63
RBM+LSTM+FNN [306]	120.54	128.70
LSTM+FNN+SC	**85.76**	**9.92**

Experimental Results

Baselines comparison. The model used here is compared with the model proposed in the literature to validate the approach. Table 8.4 presents the RMSE and SCORE with different approaches. For a fair comparison, the same RUL targets construction procedure is used. All the approaches are tested when trained on augmented data sets. For the approach with CNN, a sliding window is necessary. The window size from 20 to 80 is tested, and Table 8.4 provides the best results within the window size. It is shown that the model used in this paper provides the smallest RMSE and SCORE when compared with other approaches.

Effect of data augmentation. Figure 8.15 compares the RUL predictions on both the air filter fault and the turbo fault in the test set when the model is trained on augmented data sets and the original training set. The RUL predictions are noisier when no augmented data is used. In addition, when the model is trained on the augmented data set, the RUL prediction is close to the RUL target not only at the end of the engine's life but also at the early stage of degradation. It proves the advantage of using the data augmentation technique.

8.5 Decision Support and Decision Automation

The ultimate objective of a PHM system is to provide reliable decision support or automation to facilitate predictive maintenance scheduling and optimized logistics in the value chain. Decision support aims to assist decision-makers (DMs), such as ship engineers and/or technical superintendents on shore in making informed decisions, while decision automation employs intelligent algorithms and software to generate fully autonomous decisions [262, 307]. A typical PHM system incorporates inputs from sensor data, extensive human domain knowledge, and DMs who interpret the system outputs. However, human expert-generated input may prove inadequate when

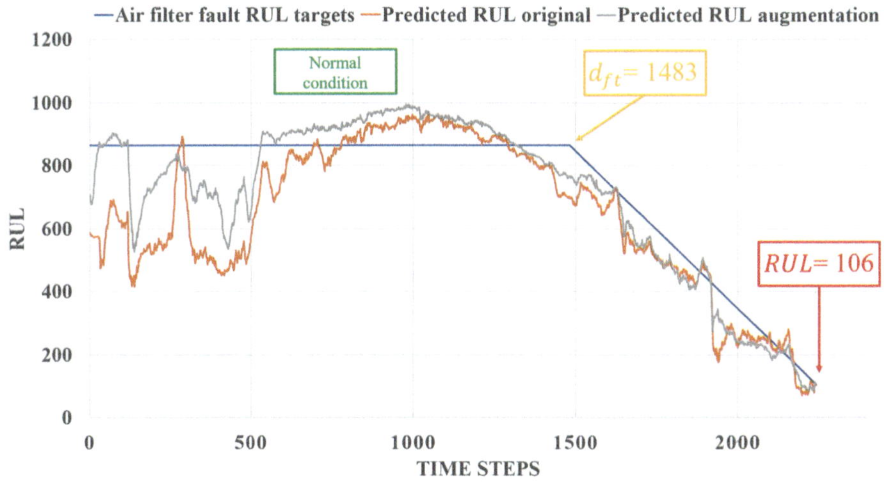

(a) Air filter fault.

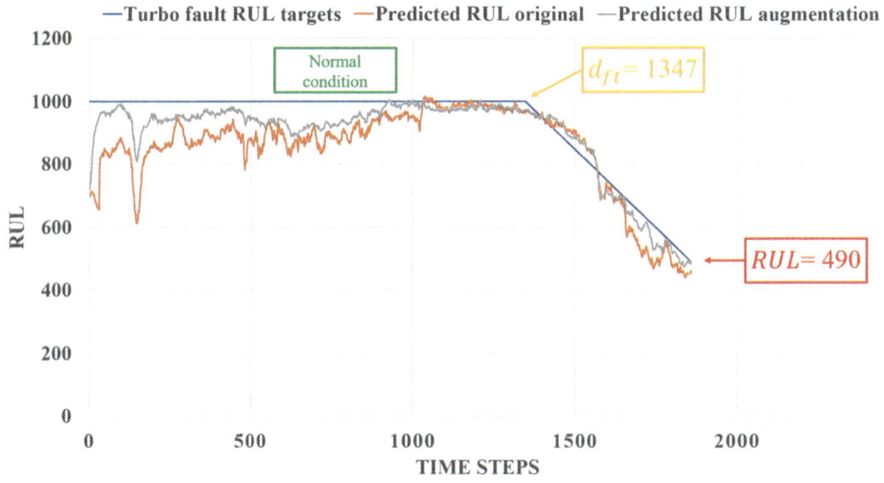

(b) Turbo fault.

Fig. 8.15 The prediction results on the air filter fault and the turbo fault in the test set

confronted with new operating conditions that fall outside the predefined knowledge base. Furthermore, even the most capable DMs may lack the cognitive capacity to effectively analyze and comprehend vast amounts of data [308]. Consequently, decision support in the era of digitalization is subject to uncertainties and does not guarantee consistently high-quality decisions. However, the neural network-based techniques, as mentioned in the above sub-chapters, enable automatic data processing and semi-supervised and unsupervised learning procedures, thereby reducing the reliance on human expert-generated input and holding the potential to enhance the intelligence of PHM systems.

According to [309], it is proposed that a fully autonomous (intelligent) PHM system must exhibit reliability exceeding 99%. This high level of reliability enables the PHM system to directly provide maintenance instructions to ship engineers and technical superintendents. Additionally, for autonomous PHM systems, it is crucial to establish their reliability to facilitate the adoption of a "black-box" approach. In instances where PHM systems have lower reliability, granting user access and explainability to the resulting analysis becomes necessary to foster understanding and build trust in the system. Additionally, the user-machine interface plays a significant role in promoting understanding and trust in a PHM system, as it greatly influences how DMs perceive and comprehend the system.

8.6 State-of-the-Art and Current Challenges in the Maritime Industry

The maritime industry has traditionally relied on reactive and preventive maintenance approaches until recent times. Reactive maintenance (RM) involves running machinery until failure occurs, leading to significant risks of unscheduled downtime. On the other hand, preventive maintenance (PM) involves adhering to predetermined intervals or running hours for maintenance, often resulting in unnecessary costs due to conservative manufacturer-recommended intervals. Consequently, functional machinery may undergo maintenance, potentially introducing failures due to human error or accelerated wear [262]. In addition to RM and PM procedures, ship engineers typically conduct oil quality and vibration analyses on gears and shafts. However, real-time indications of the machinery's actual condition are usually not available to ship engineers responsible for maintenance, as there can be a delay of one to three months between taking samples and receiving the results. In recent years, the maritime industry has witnessed a shift towards more predictive maintenance procedures, which involve integrating the anomaly detection step within a PHM system.

The utilization of sensor data for condition monitoring and the adoption of neural network-based semi-supervised anomaly detection techniques have gained commercial acceptance in the maritime industry. This approach aims to assist ship engineers in detecting or becoming aware of random faults that may occur between predetermined maintenance intervals or running hours. Semi-supervised anomaly detection serves as a natural initial step in implementing a PHM system because the training dataset primarily consists of normal sensor data, which is relatively easier to define and collect compared to anomalous and faulty data found in commercial ships. However, the majority of machinery sensors installed on modern ships are primarily intended for engine control systems, power management systems, and/or integrated alarm/automation systems (IAS), depending on the specific ship type and segment.

These sensors are typically designed to trigger immediate alarms onboard the ship, such as a pressure sensor triggering low or high alarms in response to sudden pressure increases at specific times. Consequently, these sensors are not optimized for detecting gradual degradation. As a result, semi-supervised anomaly detection may be susceptible to false negatives, as anomalous data can often resemble normal data. This further underscores the importance of possessing extensive domain knowledge in machinery to verify detected anomalies.

The main commercial objective of employing neural network-based semi-supervised anomaly detection is to detect anomalies at the earliest possible stage, ideally up to a month before an impending fault that could lead to machinery failure. This proactive approach enables ship engineers, in collaboration with technical superintendents ashore, to promptly order necessary spare parts, schedule maintenance tasks, and rectify machinery issues before potential breakdowns occur. However, when the neural network-based semi-supervised anomaly detection model is trained using data from alarm sensors on present-day ships, its design is primarily geared towards detecting alarm patterns rather than degradation patterns. Consequently, the neural network is often only capable of detecting anomalies within a short time frame before the actual fault manifests. It is evident that future ships must invest in advanced sensors during the design and construction phases to effectively measure degradation. This will facilitate the full utilization of the strengths offered by data-driven PHM methods.

8.7 Summary

In this chapter, we have explored the diverse applications of prognostic and health management (PHM) systems that offer operational benefits to ships. The concept of digital twin ships, constituting a virtual representation enabling real-time monitoring, analysis, and optimization of ship operations, has been introduced. Notably, the integration of a PHM system within digital twin ships has been highlighted as a pivotal aspect. Delving into the core components of a PHM system, namely anomaly detection, fault diagnostics, fault prognostics, decision support, and decision making, we have provided a comprehensive discussion.

Of particular focus has been the intricate formulation of anomaly detection, fault diagnostics, and fault prognostics within the framework of deep learning. Two illustrative case studies, centered on anomaly detection and fault prognostics, have been conducted to underscore the potential efficacy of employing deep learning methodologies in PHM systems tailored for the maritime industry. Furthermore, the prevailing challenges entailed in the development of PHM systems within the maritime domain have been cogently outlined, underscoring the complex terrain that necessitates diligent navigation.

While this study has showcased two illustrative case studies, a more extensive validation of the approach remains imperative to fortify its robustness. Specifically, in the context of anomaly detection, a broader dataset encompassing diverse fault types assumes paramount importance, necessitating a meticulous evaluation of the proposed method's performance through well-defined metrics like the false positive rate. For fault prognostics, the acquisition of comprehensive run-to-failure data presents a formidable challenge. To surmount this, an avenue worth exploring involves a strategic pivot towards a model-based or hybrid approach, effectively mitigating the stringent data requirements. Furthermore, a critical facet entails the meticulous identification of sensors attuned to the subtle degradation of specific ship components, thereby facilitating a more nuanced and precise prognosis.

Peihua Han received the Ph.D. degree from the Department of Ocean Operations and Civil Engineering, Norwegian University of Science and Technology (NTNU), Ålesund, Norway, in 2022. He is currently a Post-Doctoral Fellow with the Department of Ocean Operations and Civil Engineering, NTNU. His research interests include data science, deep learning, uncertain qualifications, and ocean monitoring and surveillance.

André Listou Ellefsen was awarded a Ph.D. in Engineering from the Department of Ocean Operations and Civil Engineering, within the Faculty of Engineering at the Norwegian University of Science and Technology (NTNU), Ålesund, Norway, in 2020. He is affiliated with the Intelligent Systems Laboratory at NTNU. Following the completion of his Ph.D., Dr. Ellefsen established Dipai AS, a company named as an acronym for "diagnostics and prognostics using AI", to offer a Software-as-a-Service (SaaS) solution to shipowners.

Vilmar Æsøy received the Ph.D. degree from the Norwegian University of Science and Technology (NTNU) for his research on natural gas ignition and combustion through experimental investigations and numerical simulations in 1996. Since 2002, he has been employed in teaching with the Ålesund University College, where he is also developing and teaching courses in marine product and systems design on bachelor's and master's level. In 2010, he received the green ship machinery professorship at NTNU. His current research interest includes energy and environmental technology, with focus on combustion engines and the need for more environmental friendly and energy efficient systems.

Houxiang Zhang received the Ph.D. degree in mechanical and electronic engineering from Beihang University in 2003 and the Habilitation in informatics from the University of Hamburg, Germany in 2011. He is currently a full Professor with the Department of Ocean Operations and Civil Engineering, Faculty of Engineering, Norwegian University of Science and Technology (NTNU), Norway. From 2011 to 2016, he also hold a Norwegian National GIFT Professorship on product and system design funded by the Norwegian Maritime Centre of Expertise. He has engaged into two main research areas, including control, optimization and AI application especially on autonomous vehicle; and marine automation, digitalization and ship intelligence. He has applied for and coordinated more than 30 projects supported by Norwegian Research Council (NFR), German Research Council (DFG), EU, and industry. In these areas, he has published over 300 journals and conference papers as the author or coauthor. He was nominated and selected as a member of the Norwegian Academy of Technological Sciences in 2019, and a member of Academy of the Royal Norwegian Society of Sciences and Letters in 2024.

Chapter 9
Applications in Demanding Maritime Operations

Pierre Major, Yingguang Chu, Hans Petter Hildre, and Houxiang Zhang

This chapter presents the challenges encountered when planning offshore maritime operations and how to overcome them using digital twin (DT) and virtual prototyping (VP). The first two case studies based on Gunnerus will be presented to show how DT works in real ship operations. After that, we will go even boarder to introduce how VP and advanced simulation technology enhance the demanding operation effectiveness and safety. The state-of-the-art advanced maritime operations preparation will be presented first, followed by an analysis of the level of fidelity of the component required for each step in the planning. After that, the other two case studies will be given as examples.

9.1 DT in Demanding Ship Operation

The life cycle of a ship is normally with several phases. The timeline of a ship starts typically with 6 months of design, two years of building, and 30 years in service. In the past, testing and verification have been done during the delivery of the ship. DT is a game changer to improve operations in the total lifespan of a ship. With the integration of sensors, data analytics, and artificial intelligence, DT has become a powerful tool for enhancing efficiency, productivity, and decision-making. They provide valuable insights for design improvements, predictive maintenance, and operational optimization.

9.1.1 Towards Remote Decision Support for Maritime Operations

Offshore vessels often perform in harsh environments with winds, waves, currents, and low temperatures. These operations need to be carefully planned and trained before execution. Let us investigate an example. Assume that a vessel is going to

H. Zhang et al., *Digital Twins for Vessel Life Cycle Service*,
https://doi.org/10.1007/978-981-97-4768-9_9

install a subsea factory on the seabed. The operation is usually planned and simulated a year in advance. Ship loading condition as well as weather conditions are assumed, and simulations predicting responses are implemented as follows. Ship movement responses are calculated, and maximum forces in crane wires, the risk for snap loads, DP performance and so on are simulated. In this way, the digital twin of the whole ship could be used as a valuable tool for pre-operation. DT not only allows the iterative design of operational procedures by identifying inconsistencies and blunders, exploring "what if" scenarios during onshore planning sessions based on historical Metocean records and raising awareness on critical phases, but also it can be used in "toolbox-talks" just before the execution of the operation to rehearse the procedure based on the current ship loading conditions and nowcast of Metocean conditions. Finally, DT could be used during offshore operations by visualizing information in a central place and giving insights based on physics-based simulations.

The NTNU Ålesund Ocean Møre Research Lab is a research and education facility equipped with many simulators with 3D digital twins of the environment with bathymetry, satellite pictures, height maps, and the built infrastructure such as buildings and quays. It can be used as a simulator center as well as a remote monitoring center.

Figure 9.1 illustrates the possibilities by presenting a snapshot of a remote-monitored crane operation in the Trondheim fjord, 400 km away from the monitoring center. The operation lasted about half an hour. The data from the ship and shipboard crane was received in real-time via a 4G connection. The three digital twins in Fig. 9.1 are all real-time representations of the same system consisting of the ship and crane subsystems.

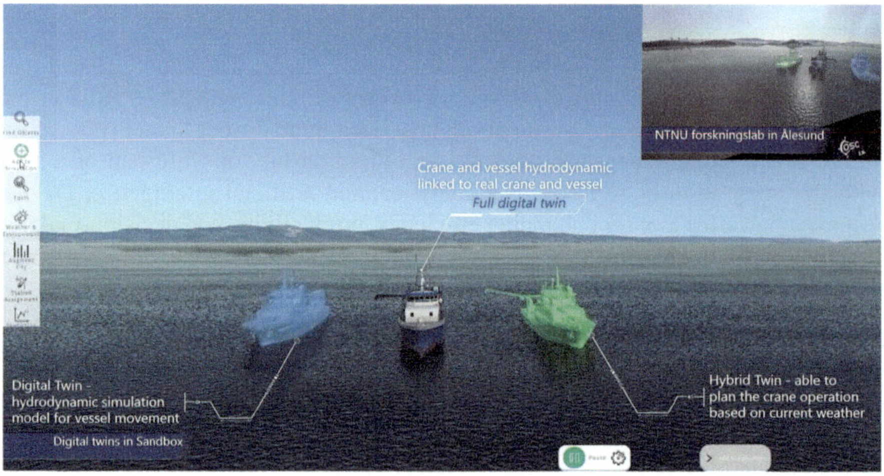

Fig. 9.1 Three real-time digital twins

- The hull movement of the transparent blue system on the left of the picture is determined by setting harsher weather in the simulation, but mirrors the real crane movement of the real vessel.
- The ship and crane with solid colors in the middle are the exact representation of the real system in real-time operation. The corresponding four channels of crane data for the whole operation, taking Fig. 4.8 as an example, can be used for crane status evaluation in the twin system.
- The hull of the transparent green ship on the right of the picture follows the movement of the real ship, but its crane is controlled by an operator in the remote centre to test alternative path trajectories in lifting operations.

All functions have been integrated into NTNU forskingslab in Ålesund based on the cooperation and support from the Offshore Simulation Centre. A real-time online DT will enhance the operation safety and security in the following aspects:

- Direct testing of "what-if" scenarios based on the real-time current state of remote systems.
- Exploring the limits of procedural safety by simulating the operation in more extreme weather conditions, with identical current crane loads, deck arrangement, and ship ballasting.
- Finding a proper way or alternative safer ways of performing the operation under certain harsh weather conditions, using short-term weather forecasts.

Figure 9.2 demonstrates the flexibility of the architecture. The crane of the green case is computed in AgX, the movement of the blue hull is computed by FhSim, and the real movement of the ship is fed in real-time through the physics abstraction layer.

The data from the ship and shipboard crane were received in real-time via a 4G connection, as detailed in Fig. 9.3. Two systems gathered the data on board the R/V Gunnerus. An OLEX server, a legacy system gathering all navigational data, was preexisting in the experiment. The onboard MQTT Broker was installed to gather a broader range of data such as engine and propulsion data. This server can be utilized to gather a wide range of data from different domains, such as data from the Crane PLC. The OLEX and MQTT servers are mirrored onshore in the research center, and the data streams flow to the instrument screen (Nav Screen) and to feed the OSC simulator's Real-Data port of Fig. 9.2 for simulation and visualization purposes.

If remotely controlled operations of ROVs and cranes are commercially available, this subsection presented an innovative case of crane and ship systems remotely monitored with humans in the loop for short-time planning purposes. Furthermore, the data logged during the operations can be used to refine the prediction models created prior to operations thus improving the planning simulations at the macro and meso planning level.

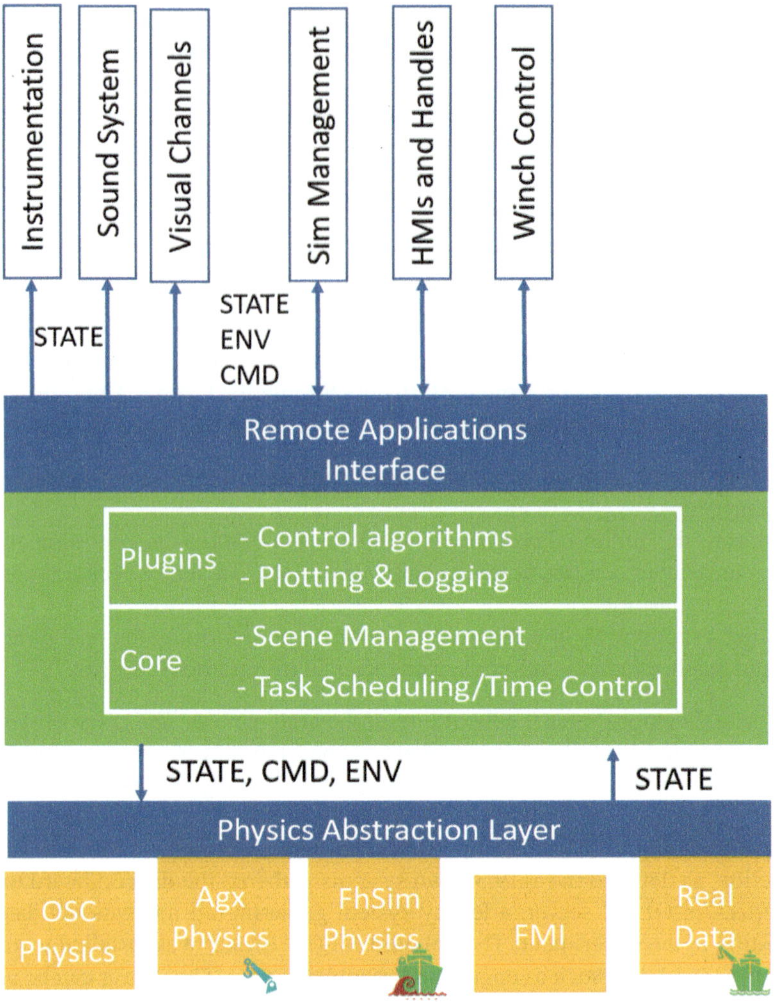

Fig. 9.2 High-level architecture Gunnerus simulator

9.1.2 Digital Twin-Based Simulation for Demanding Crane Operation

This case study will only focus on crane operation. As known, cranes are important deck equipment in many offshore and subsea applications for lifting, transferring and handling objects and personnel. Unlike land-based cranes on fixed working platforms, shipboard crane operations are affected by the ship motion in waves and the suspended load, and vice versa. Especially, large ship motions can be induced by external disturbances from heavy loading conditions in rough weather environments.

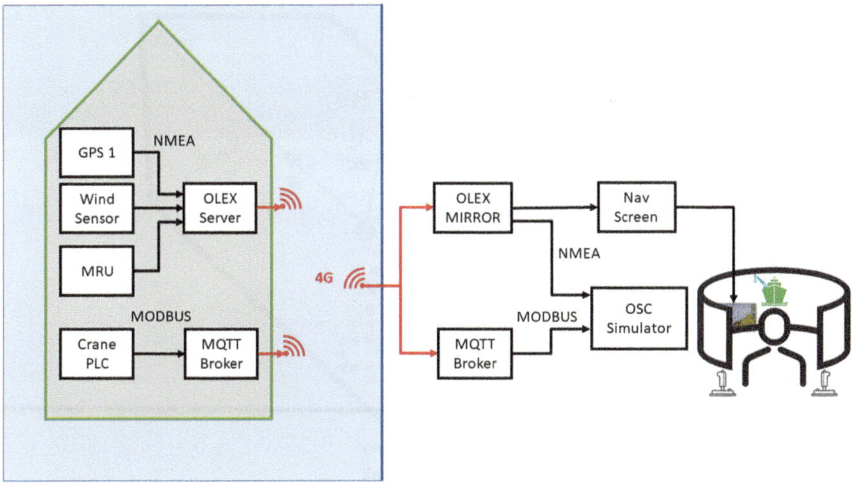

Fig. 9.3 High-level sensor network architecture Gunnerus real-time monitoring

In this case study, we present a coupled simulation of a heavy-load crane with ship motion in waves, where the dynamic forces from the crane and payload are applied to the ship.

In Chap. 5, we presented model development for the Gunnerus vessel and the Palfinger crane using object-oriented modeling (OOM), and simulation coupling using co-simulation based on the Functional Mock-up Interface (FMI) standard. The main advantage of this approach is the modularisation of complex dynamic systems for modeling and efficient coupling of domain-specific simulation tools. One of the challenges with co-simulation is the coupling of tightly-coupled systems in terms of computation. For example, a small ship and a deck crane with a heavy payload whose dynamic inertial impacts on the ship cannot be neglected during operation. Tightly-coupled systems are not naturally fit to be handled separately for co-simulation. In this very example, the crane and the payload can be connected via a flexible cable and treated as one component. The connection between the ship hull and the crane base can also be represented by a spring-damper system with high stiffness and low damping coefficient. However, this requires extremely high computing resources for real-time simulation, which is extremely critical for crane operations in this application. Alternatively, the forces from the crane including the payload are approximated as external forces and applied to variant attacking points on the ship hull during operation [112].

To couple the ship-crane-load together, transformations for the motions of the objects and reaction forces are needed. The transformation between the ship and the crane including the payload is given by the surge-sway-heave motions and roll-pitch-yaw Euler angles ψ, θ, ϕ, as shown in Fig. 9.4. Equation 9.1 gives the linear velocity transformation from the body frame to the global frame. Similarly, Eq. 9.2 gives the generalized linear force transformation. The superscript and subscript in

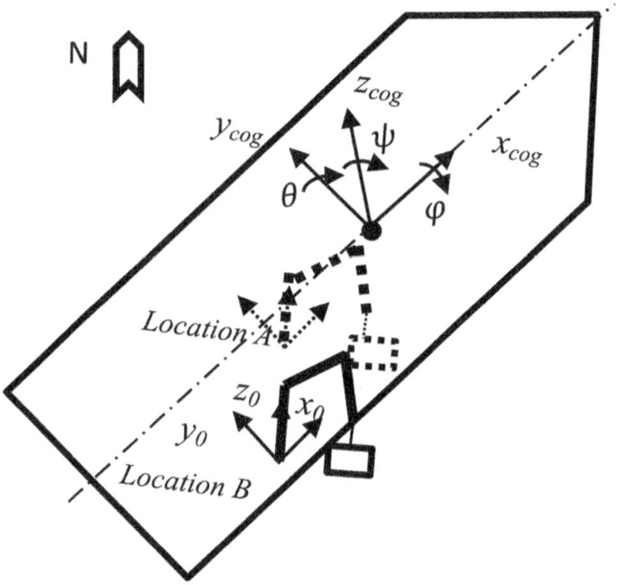

Fig. 9.4 Reference frames from ship to crane base

the equations represent the mapping between the reference frames of the system. More specifically, from the crane tip to the crane base {0}, to the center of gravity {cog} of the ship and the global frame {g}.

$$V_{tip}^{g} = V_{ship}^{g} + R(\psi, \theta, \phi)V_{tip}^{cog} \tag{9.1}$$

$$F_{ship}^{g} = V_{tip}^{g} - \sum G_i^g \tag{9.2}$$

where G_i^g denotes the gravitation force of the crane's ith link. The negative sign here represents the direction of the forces. The rotation matrix $R(\psi, \theta, \phi)$ from frame {i} to the ship frame attached to the {cog} of the ship is given by Eq. 9.3.

$$R(\psi, \theta, \phi) = \begin{bmatrix} c\psi c\theta & c\psi s\theta s\phi - s\psi c\phi & c\psi s\theta c\phi + s\psi s\phi \\ s\psi c\theta & s\psi s\theta s\phi + c\psi c\phi & s\psi s\theta c\phi - c\psi s\phi \\ -s\theta & c\theta s\phi & c\theta c\phi \end{bmatrix} \tag{9.3}$$

The lifting scenario is illustrated in Fig. 9.5. The whole operation sequence is divided into five steps. The initial state of the system is with the crane at the starboard side of the ship, with the outer boom extended over the payload.

The roll response of the ship reflects the disturbances from the crane movement, as shown in Fig. 9.6. The roll motion is excited at 50 s when the external forces are applied including the thruster forces and the crane forces. The ship gradually

Fig. 9.5 Palfinger crane lifting operation scenario

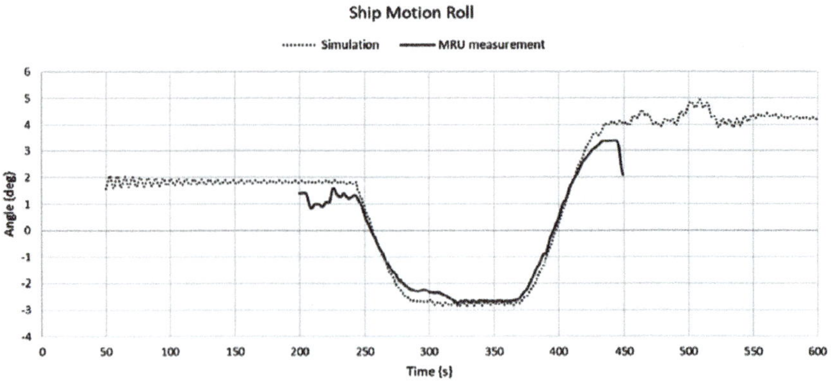

Fig. 9.6 Roll angle of the ship during crane operation in calm water

restores its position and balance after 200 s. Crane lifting operation starts from 220 s following the procedure in Fig. 9.5. The crane main boom and slewing joint angles are shown in Fig. 9.7. Small variations can be seen from the ship roll response as the main boom lifts and lowers the payload at 220 s and 300 s, respectively. The ship roll angle decreases as the slewing joint of the crane moves from the starboard side to the port side.

Figure 9.8 shows the roll response of the ship when the crane operates as described in waves. As can be seen, the roll angle increased significantly compared to calm water. The ship motions cause large movements of the crane and the payload, as shown in Fig. 9.9. As a result, large dynamic forces from the crane and payload are generated. The movements of the ship in such conditions are challenging for safe operation with a heavy load. Therefore, it is advised to start crane operation in better conditions. The load follows the crane tip movements with minimal relative offsets. The initial wire length is 1 m, which is a short segment considering the crane

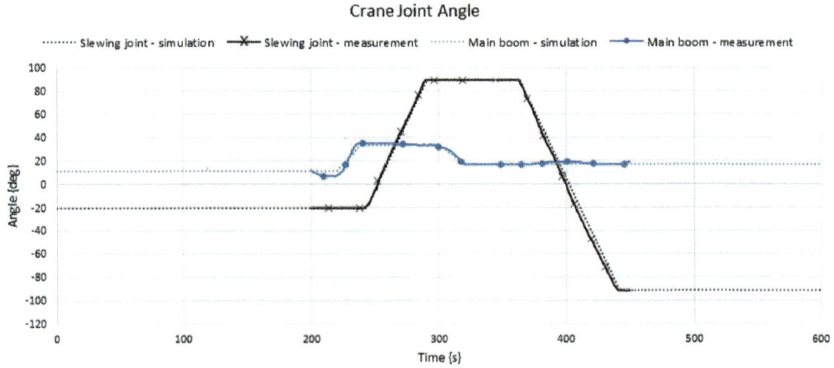

Fig. 9.7 Crane joint angles during crane operation in calm water

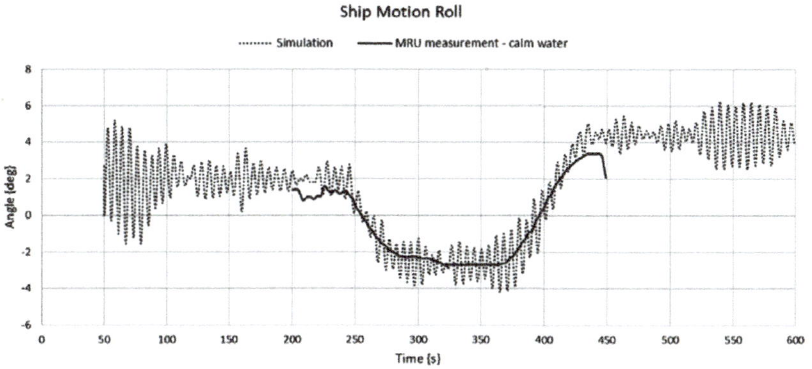

Fig. 9.8 Roll angle of the ship during crane operation in waves. Environmental conditions are: wave height 0.5 m, wave period 10 s; current velocity 1 m/s; wind velocity 3 m/s; wave, current and wind velocity direction – 135°N

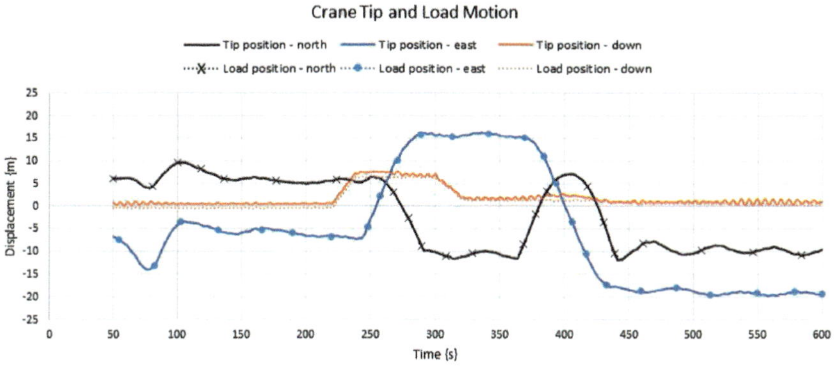

Fig. 9.9 Crane tip position and payload position during crane operation in waves. Environmental conditions are: wave height 0.3 m, wave period 10 s; current velocity 0.5 m/s; wind velocity 1 m/s; wave, current and wind velocity direction – 135°N

movement range. As a result, it shows a small load swing relative to the crane tip positions in the north and East direction. Load swing can increase significantly when the wire is released via the winch system. It would be challenging to manoeuvre the crane with a heavy load safely at such conditions without compensation.

The co-simulation approach enables the use of different domain software tools for complex multi-domain systems. It also improves the simulation efficiency through the orchestration of the sub-simulators and parallel computation using multiple processes. It depends on the application to determine the acceptable co-simulation step sizes and numerical errors. For the system presented in the test cases, the real-time index is about 30, meaning that the simulation time is about 1/30 of the clock time. However, the challenges regarding tightly-coupled systems persist. This calls for more simulation efficient solutions for both the coupled systems (i.e., the ship motions and the crane model), and the co-simulation coupling (e.g., master algorithms). Establishing the equations of motion of the ship alone has been done, while having a running model with tightly coupled system in real-time is non-trivial. Other alternative approaches to obtain the ship motion in waves can utilize data-driven techniques instead of model-based methods. In [310], a data-driven model for time series prediction of ship motion is presented. However, the performance of data-driven methods is dependent on the obtained data set for training and prediction. Such data in the maritime domain is often limited. Moreover, the accuracy of longer time prediction is another challenge for demanding operations such as crane heavy lifting. It is shown that predictions, in most cases, can be made fairly accurate up to 20 s ahead of the current time [311]. Accurate longer time predictions cannot be improved by measured motion sensor data alone.

9.2 Advanced Simulation and VP in Offshore Operations

Here we will go further to introduce the advanced simulation and VP applications in offshore operations. It is even more popular in the maritime industry since it doesn't require real-time data transmission between physical counterpart and simulation. First, it is necessary to repeat a couple of clarifications and definitions. As mentioned in Chap. 2, the term *virtual prototyping* refers to digital simulation and visualization techniques used to improve processes and products; while the *digital twins* are with strong interaction between physical counterparts and digital simulations.

Marine or Maritime operations are interventions during which the assets are installed, integrated, decommissioned, and sometimes moored. Furthermore, three planning levels need to be defined which sometimes overlap a little. Firstly, the *macro* level spans decades and has a global supply chain scope. For example, the current lifetime of a wind turbine is estimated to be 25–35 years, and the Front-End Engineering and Design (FEED) of an offshore project is generally done at this level. Engineers have a design envelope of equipment to install with a design envelope of installation methods and a wide uncertainty range of environmental parameters. Secondly, the *meso* level spans years, has a regional scope (a couple of hundreds of kilometers), and involves more detailed FEED and high-level logistics planning. In this phase, the infrastructure to be installed is decided, environmental surveys have been completed, and the installation methods and period have been decided. Finally, the *micro* level spans months to days, e.g. from the mobilization period to offshore operation completion, and has an ultra-localized spatial scope (from a couple of kilometers down to 10cm for soil mesh and 1mm precision for 3D models). At the micro level, the installation methods are decided, but weather uncertainties and installation vessel technical challenges can upset the plans from higher levels.

As known, not only for ships, but also to service offshore structures such as ships, oil rigs, wind turbines, and ocean-going fish farms, the maritime industry must plan activities in both time and space at different granularity levels: macro-, meso-, and micro-planning. The advanced simulation and virtual prototyping could provide us with an effective solution.

9.2.1 Simulating Advanced Maritime Operations

To provide engineers with solid and trustworthy facts for decision-making, for example, getting feedback from the maritime crew about the feasibility and convenience of the method statements and identifying inconsistencies and blunders, the simulations have to encompass, model the whole offshore system, and avoid overlooking critical components. As a consequence, simulations of advanced maritime operations should include the following systems, subsystems, and components:

- **Vessels**: Barges, floating vessels, semi-submersibles, and jack-up vessels often combine mooring with dynamic positioning (DP). AHTS is currently used for the towing of floaters. Service Operations Vessels, equipped with telescopic motion compensated gangways, are used for personal transfer to the wind turbines.
- **Handling systems**: Refer to cranes, cable layers, grippers or similar lifting systems with passive and active compensation systems. These systems can be electrical or hydraulic. Some have a conventional design, and some have innovative designs that need extra work to model in the simulation.
- **Power management systems (PMS)**: PMS regulates, distributes, and handles contingencies by dynamically allocating power to the respective onboard systems;
- **Ballast systems**: The system to keep the vessels stable when moving heavy loads during operations.
- **Assets and their ancillary systems**: Refer to towers, foundations, nacelles, blades, blade racks, fenders, quick connectors and hooks, sea fastenings, lifting gears, etc.
- **Waterborne equipment**: ROVs, which are tethered to the vessels and perform subsea operations such as tie-in, opening and closing valves, and guiding subsea lifts by giving camera footage.
- **Human operators**: Operators at various workstations such as DP, crane, ROV, gripper, deck, ballast, PMS, cable laying, etc.

9.2.2 Models in Virtual Prototyping for Advanced Offshore Operations

In VP of advanced offshore operations, [312] gives an overview and assessment of state-of-the-art tools and methods that can also be applied to offshore wind installation. More specifically, for example, the installation of floating offshore wind turbine is still missing a tool [313]. We coarsely classify the planning tools into three categories since there are a lot of similarities:

- **Physics model based tools**: SESAM umbrella software (SIMA/SIMO/Riflex) [314, 315] and Matlab/Simulink [316] are utilized to design method statement and perform micro planning; Orcaflex [29] is used to design mooring systems in specific offshore floating site[317].
- **Stochastic modelling tools**: Those tools for logistics [318, 319] fall outside the scope of the DT for the coarse modelling of the operations themselves, as they are based on Monte Carlo simulation and Metocean data. Assumptions must be made on how much additional time each operation phase takes for different weather scenarios. They need to be reality-checked with empirical or experimental data.
- **Machine Learning tools**: Those tools are used for asset management optimization. Authors in [320] present a series of asset management software dedicated to optimizing the overall wind farm cost by predicting component failure and corresponding planning response; they focus on health monitoring of the assets and propose cost and uptime optimizations.

Current engineering methods seldom involve human-in-the-loop simulations; human operators, such as crane drivers, captains, and remotely operated vehicle (ROV) pilots, are modeled as perfect robotic actuators performing their work. Furthermore, when the maritime workforce is involved in the simulation, it is necessary to integrate the offshore equipment and interfaces.

Chapter 3 presented the principles of co-simulation, i.e. how to simulate all the components in the same scenario but not the same simulation environment. This section will address the faithfulness requirements for each component. Critical commercial projects (see Sect. 9.2.3) require a realistic and faithful model of every relevant onboard system to mitigate the offshore risk and identify areas of improvement. But in the lack of empirical data, one has to make simplifying, if not simplistic, assumptions.

Models can be refined when the data is made available. Sometimes the component models must be simplified because of computer run time performance issues or project costs, preventing humans from being realistically in the loop. For example, if the purpose of the mission is a crane operation, one can contextualize it with plausible yet not realistic ship movement. There are, in effect, refinement levels that need to be addressed. Table 9.1 shows an assessment for each simulated component with examples of the range between basic, intermediate, and advanced. Different competencies are required throughout the system modeling: 3D artists for the modeling, hydrodynamicists for the floating structures, automation engineers for the control systems, and computer graphics and computer science engineers for the realistic and smooth real-time performance of the whole system. Experience shows that making a basic model takes a single day, and an advanced model can take many person-months equivalents.

The following section presents use cases on how the industry and academia utilize simulation and real-time monitoring to prepare maritime operations at the micro-level and how it can be feedback to the macro and meso planning.

Table 9.1 Modeling refinement levels

Module	Basic	Intermediary	Advanced
Vessel/Hull	Coarse 3D Model (no physics)	3D Model with collision and hydro-static	3D Model with collision, hydro-static, and hydrodynamics
PMS	Immediate and unlimited power for all onboard systems	Power is limited by generator max power, but no transient dynamic model	Physics-based dynamical power generation and distribution can simulate blackouts spread
DP Propulsion	Perfect Position and Orientation at all times, independent of weather and power situation	Emulated DP commands of thrusters with ship model	Real DP system fully integrated with ship model, thrusters, and PMS
Crane Hammer Gripper Handling System	Simple winches and booms are actionable with a joystick Force and moments are transferred to the vessel	Control systems entirely implemented in the physics-based model allowing testing the operational limits of the crane Human Machine Interface (HMI) is emulated on generic systems (crane/winch)	Original Equipment Manufacturer (OEM) Chair and HMI are connected to simulation System advanced operational modes are implemented (emergency, 3D or heave motion compensation, constant tension, etc.
ROV	Camera Flying in 3D, no collision	Tooling and Tether with collision Integration with ROV Chair	SIMOPS with Crane and boat and Remote operation from remote control center
Ballast	No impact of weight distribution	Weight distribution and ballast system implemented and emulated	Integration with real life ballast system interface
Environment Bathymetry	Monochromatic waves and wind, constant in time, uniform in space	Spectrum based waves and wind, scenario management allowing ramping up and down weather parameters	Hindcast and real-time weather
Cables	3D connection lines (no physics)	3D lines with realistic visuals	3D lines with realistic visuals and physics models of tension, torsion, bend, collisions, and soil friction
Assets and Ancillary systems	3D model without collision	3D model with collision and realistic weight distribution	3D model with collision and connection points to connect lifting gear

9.2.3 *Two Case Studies in Offshore Operations*

Offshore installation, maintenance and decommissioning operations involve high capital investments in ships and equipment. It often requires developing novel technological solutions and planning over many years. Due to the confidential nature of commercial operations, there is little academic literature about this industry's best practices and the state of the art. This section presents two case studies taken from two commercial projects and an in-vivo research project to show how VP and simulation could help in the operations.

Virtual Prototyping of Operations Before Equipment Delivery

Most of the heavy offshore wind operations, such as *piling*, and hammering foundations into the soil, are performed from jacked-up vessels. Jacking up and down a vessel is time-consuming and is only feasible in shallow waters. Using floating vessels with ballast is a game changer when streamlining the installation of offshore wind assets such as monopiles and offshore substations. Initially planned for delivery in 2021, DEME Offshore's flagship Orion suffered a major setback in May 2021 when a hook failure partially destroyed the 5000T heavy crane during an overload test. One year later, the DP3-certified monohull was delivered and completed more than six heavy-lift campaigns within the first summer season.

How could DEME overcome this major setback and be operational from the day of ship delivery? Parallel to the ship construction, DEME built an advanced simulator facility mirroring the ship's major control systems, much like what NASA did for their Apollo missions [321]. The simulated systems include the DP, PMS, and ballast system, gathered in the bridge, a gripper and a hammer station, and a crane dome with crane HMI shown in Fig. 9.11. To emphasize the importance of human factor in the success of operations, a debriefing room, was also delivered. This allowed the engineers and maritime crews to familiarise themselves with the various equipment components and their integration with each other more than one year before the belated ship's delivery. It also allowed DEME's engineers to design and collaboratively validate work methods for future offshore missions in the debrief room.

The high-level architecture of the distributed system is illustrated in Fig. 9.10. On the upper level, different peripheral systems are generic such as the instructor (simulation management, sound system, and visuals), or the systems specifically emulated for DEME virtual prototyping: DP, Ballast, PMS, crane chair with advanced control system, and instrumentation. The interaction between the humans and the simulation happens at this level. These peripheral applications communicate with the core of the simulation via legacy and generic remote APIs. The core synchronizes the assets' physical and environmental states (STATE and ENV), user commands (CMD), and simulation time. The physics solver Agx Physics is interfaced via an API through the physics abstraction layer and is the only one connected to the simulation.

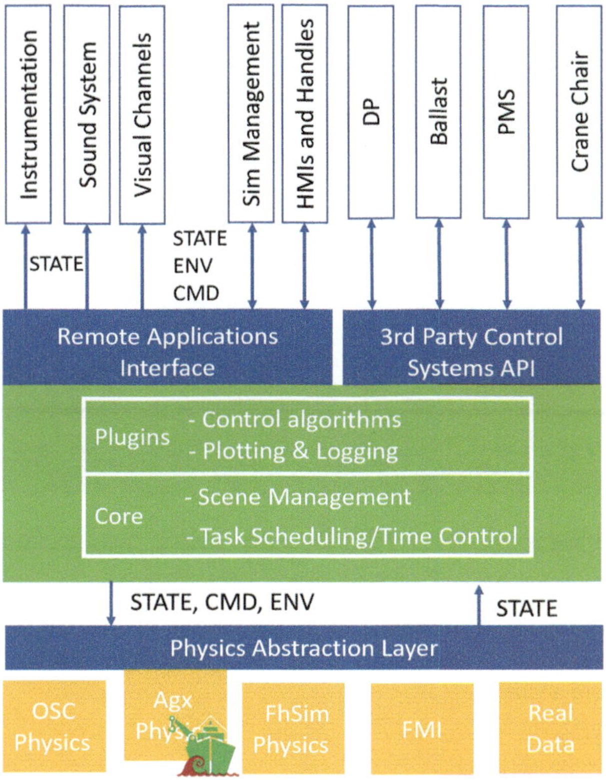

Fig. 9.10 High-Level architecture DEME simulator (courtesy of DEME)

The other physical solvers represented in Fig. 9.10, such as OSC physics (basic level physics), FhSim Physics (specific for slender hull hydrostatic and hydrodynamics), FMI (see Chap. 3), are disconnected from the simulation, as illustrated with the blank between the physics solver and the Physics Abstraction Layer. The real data is not connected to the simulation in this use case.

Engineering Procurement Construction and Installation companies(EPCIs) can mitigate risks on each mission's specifics: human factors, soil, weather, assets to be installed, deck arrangement, Jones Act-compliant feeder barges, heave, and other motion compensation systems. The simulator is usually used five to six days a week with operative crews and other experts such as the client (developers), OEM representatives (producers of wind turbine generators and ancillary systems), and Maritime Warranty Survey consultancies. Different load conditions can be imported from third-party wave analysis tools, such as WAMIT and ANSYS AQWA. The ship owner can collect the data during operations and compare it with predictions. The last VP runs are performed a couple of days before the intervention to leverage the higher accuracy of the weather forecast. This short-term feedback loop between planning and operations allows for fine-tuning of the models to improve the trustworthiness of the

Fig. 9.11 DEME crane cabin view (courtesy of DEME)

asset models and the building blocks of the simulation. Being better prepared at the micro level reduces the uncertainties at the meso level. As a result, the campaign's planning is more accurate.

Planning Tie-in Towing Simultaneous Operations

Oil and gas maritime operations are often one-off missions, with a little knowledge feedback loop between the interventions. This is due to the long intervals between the maintenance and retrofitting activities, every 5 to 15 years, or between the installation and the decommissioning, with many decades of production. For logistics and organisational reasons, different engineers and crew will perform the interventions. In other words, even if the operations have been done before, no current crew has enough operational experience or knowledge. It is thus necessary to prepare the missions by accumulating and rebuilding the knowledge and a common understanding between the stakeholders. We present here the case of Equinor, Havfram, and Global Maritime during the Njord A towing, mooring, emergency, and tie-in operations in March 2022. The focus is on the micro planning of the maritime operation involving an oil rig, towing, and subsea installation vessels.

Figure 9.12 shows a typical workflow for a simulation project with two validation sessions. Offshore operations span six to eight months from project kickoff to project completion. The modeling phases involve gathering various information such as 3D models, user manuals, load conditions, etc., and creating multi-rigid body physical models of the operation. In essence, a validation session is a simulation performed in a simulator facility with all relevant stakeholders involved based

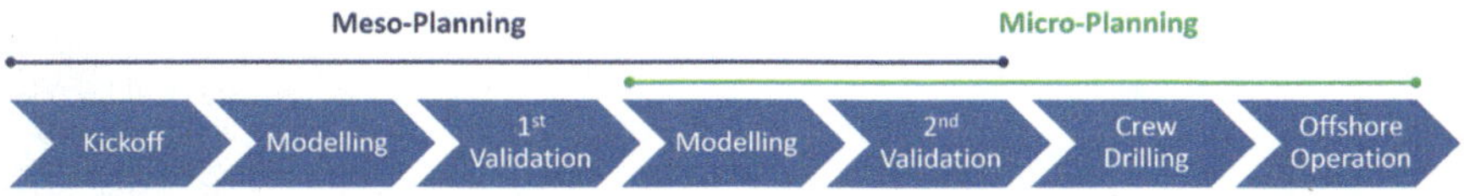

Fig. 9.12 Njord A project simulation planning

on weather conditions and other Met-Ocean statistical data for the area. Procedure blunders, interface frictions between the different contractors, and simulation calibration improvement are discovered during this first validation. A second phase of modeling and validation follows. This iteration contributes to narrowing down the design envelope. If necessary, an extra phase of modeling and validation can be added. After project engineering approval, the whole maritime crew is drilled in the simulator with life-like emergencies, with the weather forecast for the following days. This contributes to a more accurate estimation of the weather window for the operations to be undertaken. Figure 9.13 depicts the live feed views of an emergency drill during the towing of the Njord A semi-submersible platform (upper left), 3 AHTSs (lower right), with a bridge view CCTV (lower left), and a mission control room (upper right). Mariners and engineers could discuss the various coordinated responses to various emergencies.

The mooring operation of the semi-submersible is simulated. Figure 9.14 illustrates the photorealism of the simulated view (on the left), compared to the CCTV feed during operation (on the right). The similarity between the simulated and the life feeds is striking to the experts' eyes: notice the raised shark jaws, centering devices, A-frame, the tug wire, the yellow cable and the railway crane holding the end link.

Fig. 9.13 Njord A emergency training during platform simultaneous operation towing, in OSC simulator facilities, Ålesund Norway

Fig. 9.14 Mooring Operation on Njord A. Left: Simulated, courtesy of OSC. Right: CCTV feed (Courtesy of Equinor and Global Maritime)

Fig. 9.15 Simulation of riser tie-in with Njord A and multiple purpose vessel (Courtesy of Equinor and Havfram)

Finally, the tie-in operation of the oil rig to the risers involving a subsea installation vessel and its ROVs is also simulated, as depicted with transparent water in Fig. 9.15. It illustrates the process of multiple towing operations with three tugs towing the platform. The tugs are simulated in the FhSim physics solver, and the floating semi-submersible's behaviour is simulated in the Agx Physics solver. The core is responsible for the synchronization of time, position, forces, and momentum between the physics solvers. Furthermore, four different control stations are necessary to simulate all actors at play in the simultaneous operation. Having done the operations and drill in a multi-station simulator facility enhances the participants' performance and confidence during the offshore operations in real life [322]. The micro-planning of each operation amounted to better meso-planning at the campaign level by optimizing fleet utilization. Finally, the building bricks of the simulation can be reused as library assets for future operations and campaigns or real-time simulation.

9.3 Summary

In this section, we introduced how to use DT as a valuable tool for pre-operation, Two case studies on R/V Gunnerus including ship operation in different weather conditions, and crane operations are given to show its usability.

After that, we go a little bit further to general VP and advanced simulation enhancement in demanding offshore operations, not limited to ships. We briefly outlined the challenges confronting the offshore industry. We then demonstrated how industry stakeholders prepare for intricate maritime operations, highlighting instances where practical tools and practices surpass those documented in academic literature. Certain practices have the potential for enhancement through digitalization and virtual prototyping. It could help planning by creating simulation models of future systems and letting the engineers and maritime crews operate the systems prior to real life operations. If the planning simulation architecture is carefully designed, offshore operations planning at different levels can profit from the data gathered during operations, by leveraging the concept of library assets which can be fine-tuned during the life-cycle of the real assets.

Pierre Major received his M.Sc. degree in electronics and information technology from the Swiss Federal Institute of Technology of Zurich, Switzerland (ETH) in 2005, and his Ph.D. degree in Engineering from Norwegian University of Science and Technology, in 2021. He is now the Head of Research at AugmentCity in Alesund, Norway. His domains of interest are virtual prototyping of demanding offshore operations and graphical digital twins of systems such as cities or ships.

Yingguang Chu is an Associate Professor with University of Southern Denmark (SDU). Dr. Chu received his Ph.D. degree from Norwegian University of Science and Technology (NTNU) in 2018. From 2018 to 2022, he worked as a research scientist with Sintef, Norway, on digital twin system development. His current research fields and interests at SDU include mechanical engineering, hydraulic power systems, and modeling and simulation.

Hans Petter Hildre is professor and head of the Department of Ocean Operations and Civil Engineering at the Norwegian University of Science and Technology (NTNU). His area of interest is product design and system architecture design. Hans Petter is the director for Centre for Research Driven Innovation (SFI-MOVE) within marine operations. This is cooperation between NTNU, SINTEF, University Sao Paulo and 15 companies at the west coast of Norway. Professor

Hildre is head of research in national program Global Centre of Expertise Blue Maritime, project leaders in several research projects, member of the board in 5 companies, and has a number of patents.

Houxiang Zhang received the Ph.D. degree in mechanical and electronic engineering from Beihang University in 2003 and the Habilitation in informatics from the University of Hamburg, Germany in 2011. He is currently a full Professor with the Department of Ocean Operations and Civil Engineering, Faculty of Engineering, Norwegian University of Science and Technology (NTNU), Norway. From 2011 to 2016, he also hold a Norwegian National GIFT Professorship on product and system design funded by the Norwegian Maritime Centre of Expertise. He has engaged into two main research areas, including control, optimization and AI application especially on autonomous vehicle; and marine automation, digitalization and ship intelligence. He has applied for and coordinated more than 30 projects supported by Norwegian Research Council (NFR), German Research Council (DFG), EU, and industry. In these areas, he has published over 300 journals and conference papers as the author or coauthor. He was nominated and selected as a member of the Norwegian Academy of Technological Sciences in 2019, and a member of Academy of the Royal Norwegian Society of Sciences and Letters in 2024.

Chapter 10
Research Challenges and Future Perspectives

Houxiang Zhang, Guoyuan Li, and Hans Petter Hildre

This book presents an in-depth exploration of the digital twin concept applied to the research vessel Gunnerus and associated research endeavors. Comprehensive coverage is provided on key technologies such as co-simulation, component modeling, data transfer mechanisms, and tool utilization. To realize a digital twin for complex marine operations, we have embraced an architectural approach and conceptual framework utilizing an open model and simulator platform. The Gunnerus digital twin prototype is the culmination of expertise from diverse disciplines, including sensor technology, artificial intelligence, computer science, and maritime industry knowledge.

The significant contributions of our research are highlighted as follows: Firstly, we have achieved the development of a digital twin for marine vessels in operational environments, pioneering remote monitoring and control capabilities for these intricate systems. This breakthrough enables meticulous lifecycle tracking of the ship, from its design phase to retrofitting, while providing decision support for heightened safety and efficiency. The innovative work marks a significant advancement in the digitalization of real-world marine systems. While there remains ample room for improvement in accuracy and efficiency, the implementation of the Gunnerus twin ship serves as a crucial stepping stone for future developments in both maritime industry and academia.

Secondly, departing from conventional approaches that typically entail separate simulations of subsystems, we have introduced an integrated co-simulation framework for marine surface vessels. This framework seamlessly incorporates over ten multidisciplinary subsystems, addressing inherent inefficiencies and potential conflicts arising from independent simulations. The pioneering co-simulation framework, built upon the Entity-Component-System software architecture, emphasizes composability over inheritance, thus enhancing flexibility and adaptability. By concurrently simulating all systems, our approach has garnered significant interest, driving an industry-wide initiative towards the co-simulation of marine systems.

H. Zhang et al., *Digital Twins for Vessel Life Cycle Service*,
https://doi.org/10.1007/978-981-97-4768-9_10

Additionally, our research has introduced data-driven modeling into the digitalization of marine vessels, facilitating versatile decision support for various marine systems such as ship motion prediction and fault diagnostics in demanding maritime operations. Key considerations such as data dependency, imbalance, and association play pivotal roles in generating these data-driven models. Notably, we have proposed intelligent data-driven fault diagnostics and prognostics algorithms for marine diesel engines, demonstrating impressive empirical accuracy rates and generating increased interest within the maritime industry. The research outcomes have inspired the development and implementation of next-generation marine vessels and vehicles, enhancing efficiency and safety through digitalization.

The final notable contribution lies in the realm of physics-data cooperative hybrid modeling and control of marine vessels, taking into account environmental and operational uncertainties. In contrast to traditional model-based or purely data-driven approaches, there have been limited efforts to construct predictive models by seamlessly integrating both aspects in a comprehensive manner. This scarcity can be attributed to the inherent complexity of marine dynamics, exacerbated by factors such as water interaction and environmental disturbances like wind and waves. These challenges often lead to inaccuracies in traditional models of ship dynamics. To address this, our approach advocates for the integration of data-driven models as a supplementary component in ship dynamics modeling. Various methodologies were devised to accommodate differing levels of integration, contingent upon the availability and reliability of physical models. The outcomes have fostered widespread interest in refining precise ship models and advancing sophisticated control algorithms for both semi-autonomous and autonomous vessels.

Many challenges arise when trying to fully represent the physical counterpart of the twin system in a stochastic environment.

- **Developing high-fidelity models:** As learned from the Gunnerus twin ship, the amount and quality of data is fundamental to establishing the models for the twin system, e.g., either for identification of system parameters, or for development of ML models.
- **Model adaptation and trustworthiness:** Furthermore, it is very important to discuss the safe model adaptation in conjunction with model trustworthiness, in a sense of learning to model the changing dynamics of marine vessels under unknown environments, solving the persistence of changing dynamics of marine vessels in operational scenarios. Model uncertainty can be further investigated. Additionally, the model may need to provide transparent explanations of how it makes such predictions to gain the trust of the operators or industrial stakeholders.
- **Establishing a mechanism for model updating:** As twin models evolve, the system configuration frequently changes, presenting new challenges. This involves integrating new module simulation models, replacing existing ones, and adjusting model parameters. Addressing the challenge of model evolution will render the digital twin ship more promising and powerful in advancing intelligent maritime operations. It is necessary to establish mechanisms to update models, especially for these ML models regularly. Thus, the computational cost of ML models, such

as the training time for the model, can be relatively low, and the model is able to adapt to operation/environmental changes.

- **Co-simulation:** Another challenge lies in co-simulation for digital twin implementation. Scalability must be considered when it comes to the industrial use of DTs. Insufficient synchronization and tight coupling issues may arise if the system becomes complex and requires high computational effort. The domain-specific challenge in FMU implementation also needs to be addressed. Additionally, co-simulation struggles to handle tightly-coupled subsystems, such as the rigid mechanical connection between a vessel and a crane.
- **Data-driven approaches:** The lifetime of the modern vessel is about 25–30 years. During this period, the vessel can continuously collect data and an automatic label pipeline can be built. The performance of machine learning models can be improved when more data is available. From a research perspective, different learning approaches such as incremental learning [323], continual learning, or lifelong learning [324] could be considered for future research. This is extremely important for neural networks because catastrophic forgetting [325] can occur when training an already trained neural network with a new dataset. Deployed machine learning models can improve themselves over time as the ship operates.
- **Human in the loops:** The decision support interface can be designed to communicate information to vessel operators, whether onboard or at a remote control center. This will be an important step towards the availability of methods that will benefit ship operations. It could also facilitate continual or lifelong learning of the deployed models. Another related issue is how VP and DT can improve human factors issues and accelerate the planning before and during the operations.

The team at the Norwegian University of Science and Technology with our cooperated stakeholders will continue the work towards realizing a fully functional digital twin of the Gunnerus, gradually improving its accuracy. Continued development of use-cases will provide meaningful on-board decision support for the crews of the Gunnerus. The other application is digital commissioning. If we can feed different kinds of industrial data into a simulator to create a digital-twin system, and use the simulator to verify the concept, the process and the control efficiency, it will definitely reduce development time and enhance production efficiency. On the other hand, from a methodology point of view, academic research can make use of simulated data to verify the real data available, the digital-twin system can be updated and serve as a digital replica of the physical system. As a result, both industry and academia could take advantage of the digital-twin system to realize high-level applications.

Houxiang Zhang received the Ph.D. degree in mechanical and electronic engineering from Beihang University in 2003 and the Habilitation in informatics from the University of Hamburg, Germany in 2011. He is currently a full Professor with the Department of Ocean Operations and Civil Engineering, Faculty of Engineering, Norwegian University of Science and Technology (NTNU), Norway. From 2011 to 2016, he also hold a Norwegian National GIFT Professorship on product and system design funded by the Norwegian Maritime Centre of Expertise. He has engaged into two main research areas, including control, optimization and AI application especially on autonomous vehicle; and marine automation, digitalization and ship intelligence. He has applied for and coordinated more than 30 projects supported by Norwegian Research Council (NFR), German Research Council (DFG), EU, and industry. In these areas, he has published over 300 journals and conference papers as the author or coauthor. He was nominated and selected as a member of the Norwegian Academy of Technological Sciences in 2019, and a member of Academy of the Royal Norwegian Society of Sciences and Letters in 2024.

Guoyuan Li received the Ph.D. degree in computer science from the Department of Informatics, Institute of Technical Aspects of Multimodal Systems, University of Hamburg, Hamburg, Germany, in 2013. In 2014, he joined the Department of Ocean Operations and Civil Engineering, Intelligent Systems Laboratory, Norwegian University of Science and Technology (NTNU), Ålesund, Norway, where he is currently a Professor in ship intelligence. He has published more than 150 articles in the areas of his research interests, including the modeling and simulation of ship motion, autonomous navigation, intelligent control, optimization algorithms, and the locomotion control of bioinspired robots.

Hans Petter Hildre is professor and head of the Department of Ocean Operations and Civil Engineering at the Norwegian University of Science and Technology (NTNU). His area of interest is product design and system architecture design. Hans Petter is the director for Centre for Research Driven Innovation (SFI-MOVE) within marine operations. This is cooperation between NTNU, SINTEF, University Sao Paulo and 15 companies at the west coast of Norway. Professor Hildre is head of research in national program Global Centre of Expertise Blue Maritime, project leaders in several research projects, member of the board in 5 companies, and has a number of patents.

References

1. DNV, Digital twins at work in maritime and energy (2017). https://www.dnv.com/Images/DNVGLFeature2303ORIG2b_tcm8-85106.pdf, https://www.dnv.com/
2. K. Røang, Maritim21—an integrated maritime strategy for research, development and innovation. https://www.maritim21.no/contentassets/387a22cf007e4087bcbb90f16123fe1a/maritim21-presentation-english.pdf
3. B.V. Lierde, W. Cardoen, D. Radosavljevic, Predictive engineering analytics for shipbuilding—an overview, siemens PLM software, in *16th International Conference on Computer and IT Applications in the Maritime Industries*, (Hamburg, Germany, 2017), pp. 351–400, 15–17 May 2017
4. H. Feng, C. Gomes, C. Thule, K. Lausdahl, A. Iosifidis, P. G. Larsen, Introduction to digital twin engineering, in *2021 Annual Modeling and Simulation Conference (ANNSIM)* (2021), pp. 1–12
5. M. Hilbert, Big data for development: from information- to knowledge societies. Comput. Sci. Econ.: Data Collect. Data Estim. Methodol. eJournal **39**, 1–39 (2013)
6. H. Zhang, G. Li, L.I. Hatledal, Y. Chu, A.L. Ellefsen, P. Han, P. Major, R. Skulstad, T. Wang, H.P. Hildre, A digital twin of the research vessel gunnerus for lifecycle services: outlining key technologies. IEEE Robot. Autom. Mag. 2–15 (2022)
7. https://www.ntnu.edu/oceans/gunnerus
8. R. Skulstad, G. Li, T.I. Fossen, B. Vik, H. Zhang, A hybrid approach to motion prediction for ship docking-integration of a neural network model into the ship dynamic model. IEEE Trans. Instrum. Meas. **70**, 1–11 (2021)
9. Y. Chu, Y. Deng, B.S. Pedersen, H. Zhang, Parameterization and visualization of marine crane concept design (2016)
10. B.S. Pedersen, Y. Chu, H. Zhang, Virtual prototyping for maritime winch design and operations based on functional mock-up interface co-simulation. Ships Offshore Struct. **14**(sup1), 261–269
11. T. Wang, R. Skulstad, M. Kanazawa, G. Li, V. Æsøy, H. Zhang, Physics-informed data-driven approach for ship docking prediction, in *2022 IEEE International Conference on Real-time Computing and Robotics (RCAR)* (2022), pp. 111–117
12. B. Wu, M.L. Sæter, H.P. Hildre, H. Zhang, G. Li, Experiment design and implementation for human-in-the-loop study towards maritime autonomous surface ships, in *2022 IEEE International Conference on Systems, Man, and Cybernetics (SMC)* (IEEE, 2022), pp. 1948–1953

H. Zhang et al., *Digital Twins for Vessel Life Cycle Service*,
https://doi.org/10.1007/978-981-97-4768-9

13. Z. Liu, Y. Chu, G. Li, H.P. Hildre, H. Zhang, A co-simulation approach to onboard support of marine operation: a palfinger crane path planning case. Simulation, 00375497241228623
14. A.L. Ellefsen, E. Bjørlykhaug, V. Æsøy, H. Zhang, An unsupervised reconstruction-based fault detection algorithm for maritime components. IEEE Access **7**, 16101–16109 (2019)
15. R. Skulstad, G. Li, T.I. Fossen, B. Vik, H. Zhang, Dead reckoning of dynamically positioned ships: using an efficient recurrent neural network. IEEE Robot. Autom. Mag. **26**(3), 39–51 (2019)
16. L.I. Hatledal, Y. Chu, A. Styve, H. Zhang, Vico: an entity-component-system based co-simulation framework. Simul. Model. Pract. Theory **108**, 102243 (2021)
17. A. Rasheed, O. San, T. Kvamsdal, Digital twin: values, challenges and enablers from a modeling perspective. IEEE Access **8**, 21980–22012 (2020)
18. L.I. Hatledal, R. Skulstad, G. Li, A. Styve, H. Zhang, Co-simulation as a fundamental technology for twin ships. MIC—Model. Identif. Control **41**(4), 297–311 (2020)
19. F. Perabo, D. Park, M.K. Zadeh, Ø. Smogeli, L. Jamt, Digital twin modelling of ship power and propulsion systems: application of the open simulation platform (OSP), in *2020 IEEE 29th International Symposium on Industrial Electronics (ISIE)* (2020), pp. 1265–1270
20. W. Kritzinger, M. Karner, G. Traar, J. Henjes, W. Sihn, Digital twin in manufacturing: a categorical literature review and classification, IFAC-PapersOnLine **51**(11), 1016–1022 (2018). *16th IFAC Symposium on Information Control Problems in Manufacturing INCOM 2018*
21. M. Safeea, P. Neto, KUKA sunrise toolbox: interfacing collaborative robots with MATLAB. IEEE Robot. Autom. Mag. **26**(1), 91–96 (2019)
22. W. Chengbin Q. Long, Study on simulation and experiment of hydraulic excavator's work device based on simulation x, in *2011 International Conference on Electric Information and Control Engineering* (2011), pp. 1742–1745
23. A. Guironnet, M. Saugier, S. Petitrenaud, F. Xavier, P. Panciatici, Towards an open-source solution using Modelica for time-domain simulation of power systems, in *2018 IEEE PES Innovative Smart Grid Technologies Conference Europe (ISGT-Europe)* (2018), pp. 1–6
24. Y. Chu, V. Æsøy, S. Ehlers, H. Zhang, Integrated multi-domain system modelling and simulation for offshore crane operations. Ship Technol. Res. **62**(1), 36–46 (2015)
25. R. Pryor, *Multiphysics Modeling Using COMSOL 5 and MATLAB* (Mercury Learning and Information, 2021)
26. M. Servin, M. Brandl, *Physics-based virtual environments for autonomous earthmoving and mining machinery, in Commercial Vehicle Technology 2018* (Springer, Berlin, 2018), pp. 493–504
27. M. Hannan, A. Mohamed, PSCAD/EMTDC simulation of unified series-shunt compensator for power quality improvement. IEEE Trans. Power Deliv. **20**(2), 1650–1656 (2005)
28. J. Zhang, J. Zhong, R. Gao, L. He, The application of co-simulation technology of ANSYS and MSC.ADAMS in structural engineering, in *2010 International Conference on Mechanic Automation and Control Engineering* (2010), pp. 1033–1036
29. J. Jia, J. Gu, L. Chen, L. Gao, S. Wang, A numerical study on the installation configuration design of compliant vertical access riser (CVAR) in deepwater. Mar. Struct. **87**, 103330 (2023)
30. P. Major, S. Yuan, H. Zhang, I. Blaalid, M. Edet, M. Ferstad, Flexible riser installation optimisation based on virtual prototyping, in *OCEANS 2019—Marseille* (2019), pp. 1–6
31. X. Shao, J. Zhang, Z. Zhao, The position servo system based on dSPACE, in *2012 24th Chinese Control and Decision Conference (CCDC)* (2012), pp. 2515–2519
32. X. Zhao, K. Luo, The use of MATLAB and GT-suite in simulation and optimization of the diesel exhaust after-treatment system, in *2018 International Conference on Security, Pattern Analysis, and Cybernetics (SPAC)* (2018), pp. 264–267
33. T. Blochwitz, M. Otter, J. Åkesson, M. Arnold, C. Clauss, H. Elmqvist, M. Friedrich, A. Junghanns, J. Mauss, D. Neumerkel, H. Olsson, A. Viel, Functional mockup interface 2.0: the standard for tool independent exchange of simulation models, in *Proceedings 9th International Modelica Conference* (2012), pp. 173–184, The Modelica Association, See also https://www.fmi-standard.org/

34. J. Rumbaugh, I. Jacobson, G. Booch, *Unified Modeling Language Reference Manual*, 2nd edn. (Pearson Higher Education, Boston, MA, 2004), pp. 3–11
35. S. Friedenthal, A. Moore, R. Steiner, *A Practical Guide to SysML: The Systems Modeling Language* (Morgan Kaufmann, Waltham, MA, 2014), pp. 29–49
36. N. Amálio, R. Payne, A. Cavalcanti, J. Woodcock, Checking SysML models for Co-simulation BT—formal methods and software engineering, in *Proceedings of the 18th International Conference on Formal Engineering Methods, ICFEM Tokyo, Japan* (Springer International Publishing, Cham, 2016), pp. 450–465
37. A.L. Ramos, J.V. Ferreira, J. Barceló, Model-based systems engineering: an emerging approach for modern systems. IEEE Transactions on Syst., Man, Cybern., Part C: Appl. Rev. **42**(1), 101–111 (2012)
38. V. Hassani, A. Ross, Ø. Selvik, D. Fathi, F. Sprenger, T.E. Berg, Time domain simulation model for research vessel gunnerus, vol. 7, in *Ocean Engineering of International Conference on Offshore Mechanics and Arctic Engineering* (2015). V007T06A013
39. S. Skjong, M. Rindarøy, L.T. Kyllingstad, V. Æsøy, E. Pedersen, Virtual prototyping of maritime systems and operations: applications of distributed co-simulations. J. Mar. Sci. Technol. **23**(4), 835–853 (2018)
40. Y. Chu, G. Li, H. Zhang, Incorporation of ship motion prediction into active heave compensation for offshore crane operation, in *2020 15th IEEE Conference on Industrial Electronics and Applications (ICIEA)* (2020), pp. 1444–1449
41. P. Major, R. Zghyer, H. Zhang, H.P. Hildre, A framework for rapid virtual prototyping: a case study with the Gunnerus research vessel. Ship Technol. Res. **70**(1), 1–13 (2023)
42. Y. Chu, L.I. Hatledal, H. Zhang, V. Æsøy, S. Ehlers, Virtual prototyping for maritime crane design and operations. J. Mar. Sci. Technol. **23**(4), 754–766 (2018)
43. P. Major, *Virtual Prototyping for Mission Planning of Demanding Maritime Operations*. Ph.D. thesis, Department of Ocean Operations and Civil Engineering, Norwegian University of Science and Technology, Aalesund, Norway (2021)
44. G.G. Wang, Definition and Review of Virtual Prototyping. J. Comput. Inf. Sci. Eng. **2**, 232–236 (2003)
45. P. Major, H. Zhang, H.P. Hildre, M. Edet, Virtual prototyping of offshore operations: a review. Ship Technol. Res. **68**(2), 84–101 (2021)
46. R. Martinez, S. Arnau, C. Scullion, P. Collins, R.D. Neilson, M. Kapitaniak, Virtual prototyping of a floating wind farm anchor during underwater towing operations, in *Recent Trends in Wave Mechanics and Vibrations*. ed. by Z. Dimitrovová, P. Biswas, R. Gonçalves, T. Silva (Springer International Publishing, Cham, 2023), pp. 555–564
47. X. Zhang, M. Duan, D. Mao, Y. Yu, J. Yu, Y. Wang, A mathematical model of virtual simulation for deepwater installation of subsea production facilities. Ships Offshore Struct. **12**(2), 182–195 (2017)
48. Y. Yu, M. Duan, C. Sun, Z. Zhong, H. Liu, A virtual reality simulation for coordination and interaction based on dynamics calculation. Ships Offshore Struct. **12**(6), 873–884 (2017)
49. S. Stoumpos, G. Theotokatos, C. Mavrelos, E. Boulougouris, Towards marine dual fuel engines digital twins-integrated modelling of thermodynamic processes and control system functions. J. Mar. Sci. Eng. **8**(3) (2020)
50. S.-W. Ji, M.-S. Choi, Y.-B. Kim, A study on position mooring system design for the vessel moored by mooring lines. IEEE/ASME Trans. Mechatron. **20**(6), 2824–2831 (2015)
51. Y. Sha, J. Amdahl, A. Aalberg, Z. Yu, Numerical investigations of the dynamic response of a floating bridge under environmental loadings. Ships Offshore Struct. **13**(sup1), 113–126 (2018)
52. G. Li, P.B. Skogeng, Y. Deng, L.I. Hatledal, H. Zhang, Towards a virtual prototyping framework for ship maneuvering in offshore operations, in *OCEANS 2016—Shanghai* (2016), pp. 1–6
53. S. Yuan, P. Major, H. Zhang, Flexible riser replacement operation based on advanced virtual prototyping. Ocean Eng. **210**, 107502 (2020)

54. P. Major, R. Skulstad, G. Li, H. Zhang, Virtual prototyping: a case study of positioning systems for drilling operations in the Barents sea. Ships Offshore Struct. **14**(sup1), 364–373 (2019)
55. Y. Chu, L.I. Hatledal, V. Æsøy, S. Ehlers, H. Zhang, An object-oriented modeling approach to virtual prototyping of marine operation systems based on functional mock-up interface co-simulation. J. Offshore Mech. Arct. Eng. **140**(11), 021601 (2017)
56. B. He, K.-J. Bai, Digital twin-based sustainable intelligent manufacturing: a review. Adv. Manuf. **9**, 1–21 (2021)
57. F. Tao, H. Zhang, A. Liu, A.Y. Nee, Digital twin in industry: state-of-the-art. IEEE Trans. Ind. Inform. **15**(4), 2405–2415 (2018)
58. W. Kritzinger, M. Karner, G. Traar, J. Henjes, W. Sihn, Digital twin in manufacturing: a categorical literature review and classification. IFAC-PapersOnline **51**(11), 1016–1022 (2018)
59. C. Cimino, E. Negri, L. Fumagalli, Review of digital twin applications in manufacturing. Comput. Ind. **113**, 103130 (2019)
60. G. Zhou, C. Zhang, Z. Li, K. Ding, C. Wang, Knowledge-driven digital twin manufacturing cell towards intelligent manufacturing. Int. J. Prod. Res. **58**(4), 1034–1051 (2020)
61. F. Tao, M. Zhang, Digital twin shop-floor: a new shop-floor paradigm towards smart manufacturing. IEEE Access **5**, 20418–20427 (2017)
62. Q. Qi, F. Tao, Y. Zuo, D. Zhao, Digital twin service towards smart manufacturing. Procedia Cirp **72**, 237–242 (2018)
63. W. Polini, A. Corrado, Digital twin of composite assembly manufacturing process. Int. J. Prod. Res. **58**(17), 5238–5252 (2020)
64. J. Bao, D. Guo, J. Li, J. Zhang, The modelling and operations for the digital twin in the context of manufacturing. Enterp. Inf. Syst. **13**(4), 534–556 (2019)
65. P. Aivaliotis, K. Georgoulias, G. Chryssolouris, The use of digital twin for predictive maintenance in manufacturing. Int. J. Comput. Integr. Manuf. **32**(11), 1067–1080 (2019)
66. D. Piromalis, A. Kantaros, Digital twins in the automotive industry: the road toward physical-digital convergence. Appl. Syst. Innov. **5**(4), 65 (2022)
67. S. Gallego-García, J. Reschke, M. García-García, Design and simulation of a capacity management model using a digital twin approach based on the viable system model: case study of an automotive plant. Appl. Sci. **9**(24), 5567 (2019)
68. F. Caputo, A. Greco, M. Fera, R. Macchiaroli, Digital twins to enhance the integration of ergonomics in the workplace design. Int. J. Ind. Ergon. **71**, 20–31 (2019)
69. H. Xiong, Z. Wang, G. Wu, Y. Pan, Design and implementation of digital twin-assisted simulation method for autonomous vehicle in car-following scenario. J. Sens. **2022** (2022)
70. S. Baskaran, F.A. Niaki, M. Tomaszewski, J.S. Gill, Y. Chen, Y. Jia, L. Mears, V. Krovi, Digital human and robot simulation in automotive assembly using siemens process simulate: a feasibility study. Procedia Manuf. **34**, 986–994 (2019)
71. N. Kousi, C. Gkournelos, S. Aivaliotis, C. Giannoulis, G. Michalos, S. Makris, Digital twin for adaptation of robots' behavior in flexible robotic assembly lines. Procedia Manuf. **28**, 121–126 (2019)
72. X. Sun, R. Zhang, S. Liu, Q. Lv, J. Bao, J. Li, A digital twin-driven human–robot collaborative assembly-commissioning method for complex products. Int. J. Adv. Manuf. Technol. 1–14 (2021)
73. B.A. Talkhestani, N. Jazdi, W. Schloegl, M. Weyrich, Consistency check to synchronize the digital twin of manufacturing automation based on anchor points. Procedia CIRP **72**, 159–164 (2018)
74. R. Bambura, M. Šolc, M. Dado, L. Kotek, Implementation of digital twin for engine block manufacturing processes. Appl. Sci. **10**(18), 6578 (2020)
75. J.M. Rožanec, J. Lu, J. Rupnik, M. Škrjanc, D. Mladenić, B. Fortuna, X. Zheng, D. Kiritsis, Actionable cognitive twins for decision making in manufacturing. Int. J. Prod. Res. **60**(2), 452–478 (2022)
76. R. Magargle, L. Johnson, P. Mandloi, P. Davoudabadi, O. Kesarkar, S. Krishnaswamy, J. Batteh, A. Pitchaikani, A simulation-based digital twin for model-driven health monitoring and

predictive maintenance of an automotive braking system, in *Proceedings of the 12th International Modelica Conference, Prague, Czech Republic, 15–17 May 2017*, no. 132 (Linköping University Electronic Press, 2017) pp. 35–46
77. P. Rajesh, N. Manikandan, C. Ramshankar, T. Vishwanathan, C. Sathishkumar, Digital twin of an automotive brake pad for predictive maintenance. Procedia Comput. Sci. **165**, 18–24 (2019)
78. S. Venkatesan, K. Manickavasagam, N. Tengenkai, N. Vijayalakshmi, Health monitoring and prognosis of electric vehicle motor using intelligent-digital twin. IET Electr. Power Appl. **13**(9), 1328–1335 (2019)
79. G. Xie, K. Yang, C. Xu, R. Li, S. Hu, Digital twinning based adaptive development environment for automotive cyber-physical systems. IEEE Trans. Ind. Inf. **18**(2), 1387–1396 (2021)
80. G. Bhatti, H. Mohan, R.R. Singh, Towards the future of smart electric vehicles: digital twin technology. Renew. Sustain. Energy Rev. **141**, 110801 (2021)
81. Z. Wang, X. Liao, X. Zhao, K. Han, P. Tiwari, M. J. Barth, G. Wu, A digital twin paradigm: vehicle-to-cloud based advanced driver assistance systems, in *2020 IEEE 91st Vehicular Technology Conference (VTC2020-Spring)* (IEEE, 2020), pp. 1–6
82. C.M. Ezhilarasu, Z. Skaf, I.K. Jennions, Understanding the role of a digital twin in integrated vehicle health management (IVHM), in *2019 IEEE International Conference on Systems, Man and Cybernetics (SMC)* (IEEE, 2019), pp. 1484–1491
83. B. Wu, W.D. Widanage, S. Yang, X. Liu, Battery digital twins: perspectives on the fusion of models, data and artificial intelligence for smart battery management systems. Energy AI **1**, 100016 (2020)
84. M. Ibrahim, A. Rassõlkin, T. Vaimann, A. Kallaste, Overview on digital twin for autonomous electrical vehicles propulsion drive system. Sustainability **14**(2), 601 (2022)
85. Z. Hu, S. Lou, Y. Xing, X. Wang, D. Cao, C. Lv, *Review and perspectives on driver digital twin and its enabling technologies for intelligent vehicles* (IEEE Trans. Intell, Veh, 2022)
86. G. Bachelor, E. Brusa, D. Ferretto, A. Mitschke, Model-based design of complex aeronautical systems through digital twin and thread concepts. IEEE Syst. J. **14**(2), 1568–1579 (2019)
87. P. Winkler, S. Gallego-García, M. Groten, Design and simulation of a digital twin mobility concept: an electric aviation system dynamics case study with capacity constraints. Appl. Sci. **12**(2), 848 (2022)
88. B. Liang, W. Liu, K. Liu, M. Zhou, Y. Zhang, Z. Jia, A displacement field perception method for component digital twin in aircraft assembly. Sensors **20**(18), 5161 (2020)
89. X. Chang, X. Jia, K. Liu, H. Hu, Knowledge-enabled digital twin for smart designing of aircraft assembly line. Assem. Autom. **41**(4), 441–456 (2021)
90. F. Mhenni, F. Vitolo, A. Rega, R. Plateaux, P. Hehenberger, S. Patalano, J.-Y. Choley, Heterogeneous models integration for safety critical mechatronic systems and related digital twin definition: application to a collaborative workplace for aircraft assembly. Appl. Sci. **12**(6), 2787 (2022)
91. C. Mandolla, A.M. Petruzzelli, G. Percoco, A. Urbinati, Building a digital twin for additive manufacturing through the exploitation of blockchain: a case analysis of the aircraft industry. Comput. Ind. **109**, 134–152 (2019)
92. S. Liu, J. Bao, Y. Lu, J. Li, S. Lu, X. Sun, Digital twin modeling method based on biomimicry for machining aerospace components. J. Manuf. Syst. **58**, 180–195 (2021)
93. Z. Xu, F. Ji, S. Ding, Y. Zhao, Y. Zhou, Q. Zhang, F. Du, Digital twin-driven optimization of gas exchange system of 2-stroke heavy fuel aircraft engine. J. Manuf. Syst. **58**, 132–145 (2021)
94. S. Utzig, R. Kaps, S.M. Azeem, A. Gerndt, Augmented reality for remote collaboration in aircraft maintenance tasks, in *2019 IEEE Aerospace Conference* (IEEE, 2019), pp. 1–10
95. M. Xiong, H. Wang, Q. Fu, Y. Xu, Digital twin-driven aero-engine intelligent predictive maintenance. Int. J. Adv. Manuf. Technol. **114**(11–12), 3751–3761 (2021)
96. F. Hu, Digital twin-driven reconfigurable fixturing optimization for trimming operation of aircraft skins. Aerospace **9**(3), 154 (2022)

97. C.M. Ezhilarasu, Z. Skaf, I.K. Jennions, A generalised methodology for the diagnosis of aircraft systems. IEEE Access **9**, 11437–11454 (2021)
98. E. Giannaros, A. Kotzakolios, V. Kostopoulos, G. Sotiriadis, R. Vignjevic, N. Djordjevic, M. Boccaccio, M. Meo, Low-and high-fidelity modeling of sandwich-structured composite response to bird strike, as tools for a digital-twin-assisted damage diagnosis. Int. J. Impact Eng. **160**, 104058 (2022)
99. X. Zhou, S. He, L. Dong, S.N. Atluri, Real-time prediction of probabilistic crack growth with a helicopter component digital twin. AIAA J. **60**(4), 2555–2567 (2022)
100. C. Li, S. Mahadevan, Y. Ling, S. Choze, L. Wang, Dynamic Bayesian network for aircraft wing health monitoring digital twin. AIAA J. **55**(3), 930–941 (2017)
101. M. Xiong, H. Wang, Digital twin applications in aviation industry: a review. Int. J. Adv. Manuf. Technol. **121**(9–10), 5677–5692 (2022)
102. S. Yoon, S. Lee, S. Kye, I.-H. Kim, H.-J. Jung, B.F. Spencer Jr., Seismic fragility analysis of deteriorated bridge structures employing a UAV inspection-based updated digital twin. Struct. Multidiscip. Optim. **65**(12), 346 (2022)
103. N.M. Levine, B.F. Spencer Jr., Post-earthquake building evaluation using UAVs: A BIM-based digital twin framework. Sensors **22**(3), 873 (2022)
104. T. Deng, K. Zhang, Z.-J.M. Shen, A systematic review of a digital twin city: a new pattern of urban governance toward smart cities. J. Manag. Sci. Eng. **6**(2), 125–134 (2021)
105. L. Nikolopoulos, E. Boulougouris, A novel method for the holistic, simulation driven ship design optimization under uncertainty in the big data era. Ocean Eng. **218**, 107634 (2020)
106. M.S. Sapkota, E. Apeh, M. Hadfield, R. Haratian, R. Adey, J. Baynham, An approach for adaptive model performance validation within digital twinning. Int. J. Comput. Methods Exp. Meas. **9**(3), 213–225 (2021)
107. K.N. Anyfantis, An abstract approach toward the structural digital twin of ship hulls: a numerical study applied to a box girder geometry. Proc. Inst. Mech. Eng., Part M: J. Eng. Marit. Environ. **235**(3), 718–736 (2021)
108. L. Li, D. Liu, J. Liu, H.-G. Zhou, J. Zhou, et al., Quality prediction and control of assembly and welding process for ship group product based on digital twin. Scanning **2020** (2020)
109. Q. Wu, Y. Mao, J. Chen, C. Wang, Application research of digital twin-driven ship intelligent manufacturing system: pipe machining production line. J. Mar. Sci. Eng. **9**(3), 338 (2021)
110. Z. Hu, X. Fang, J. Zhang, A digital twin-based framework of manufacturing workshop for marine diesel engine. Int. J. Adv. Manuf. Technol. **117**(11–12), 3323–3342 (2021)
111. L.I. Hatledal, R. Skulstad, G. Li, A. Styve, H. Zhang, Co-simulation as a fundamental technology for twin ships (2020)
112. Y. Chu, G. Li, L.I. Hatledal, F.T. Holmeset, H. Zhang, Coupling of dynamic reaction forces of a heavy load crane and ship motion responses in waves. Ships Offshore Struct. **16**(sup1), 58–67 (2021)
113. J.-H. Lee, Y.-S. Nam, Y. Kim, Y. Liu, J. Lee, H. Yang, Real-time digital twin for ship operation in waves. Ocean Eng. **266**, 112867 (2022)
114. A. Coraddu, L. Oneto, D. Ilardi, S. Stoumpos, G. Theotokatos, Marine dual fuel engines monitoring in the wild through weakly supervised data analytics. Eng. Appl. Artif. Intell. **100**, 104179 (2021)
115. S. Stoumpos, G. Theotokatos, A novel methodology for marine dual fuel engines sensors diagnostics and health management. Int. J. Engine Res. **23**(6), 974–994 (2022)
116. E. VanDerHorn, Z. Wang, S. Mahadevan, Towards a digital twin approach for vessel-specific fatigue damage monitoring and prognosis. Reliab. Eng. Syst. Saf. **219**, 108222 (2022)
117. M. Ibrion, N. Paltrinieri, A.R. Nejad, On risk of digital twin implementation in marine industry: larning from aviation industry. J. Phys.: Conf. Ser. **1357**, 012009 (IOP Publishing, 2019)
118. V. Arrichiello, P. Gualeni, Systems engineering and digital twin: a vision for the future of cruise ships design, production and operations. Int. J. Interact. Des. Manuf. (IJIDeM) **14**, 115–122 (2020)
119. H. Zhang, G. Li, L.I. Hatledal, Y. Chu, A.L. Ellefsen, P. Han, P. Major, R. Skulstad, T. Wang, H.P. Hildre, A digital twin of the research vessel Gunnerus for lifecycle services: outlining key technologies. IEEE Robot. Autom. Mag. **30**(3), 6–19 (2023)

120. N. Assani, P. Matić, M. Katalinić, Ship's digital twin-a review of modelling challenges and applications. Appl. Sci. **12**(12), 6039 (2022)
121. D.C. Mombiela, M. Zadeh, Integrated design and control approach for marine power systems based on operational data; digital twin to design, in *2021 IEEE Transportation Electrification Conference & Expo (ITEC)* (IEEE, 2021), pp. 520–527
122. D.G. George, K.D. Eleftherios, G.A. Chariklia, LNG carrier two-stroke propulsion systems: a comparative study of state of the art reliquefaction technologies. Energy **195**, 116997 (2020)
123. Y. Liu, H. Ren, Acquisition method of evaluation stress for the digital twin model of ship monitoring structure. Appl. Ocean Res. **129**, 103368 (2022)
124. D.-J. Cheng, J. Zhang, Z.-T. Hu, S.-H. Xu, X.-F. Fang, A digital twin-driven approach for on-line controlling quality of marine diesel engine critical parts. Int. J. Precis. Eng. Manuf. **21**, 1821–1841 (2020)
125. R.E. Nielsen, D. Papageorgiou, L. Nalpantidis, B.T. Jensen, M. Blanke, Machine learning enhancement of manoeuvring prediction for ship digital twin using full-scale recordings. Ocean Eng. **257**, 111579 (2022)
126. F. Mauro, A. Kana, Digital twin for ship life-cycle: a critical systematic review. Ocean Eng. **269**, 113479 (2023)
127. X. Fang, H. Wang, W. Li, G. Liu, B. Cai, Fatigue crack growth prediction method for offshore platform based on digital twin. Ocean Eng. **244**, 110320 (2022)
128. C. Kim, M.-C. Dinh, H.-J. Sung, K.-H. Kim, J.-H. Choi, L. Graber, I.-K. Yu, M. Park, Design, implementation, and evaluation of an output prediction model of the 10 mw floating offshore wind turbine for a digital twin. Energies **15**(17), 6329 (2022)
129. A. Lambertini, M. Menghini, J. Cimini, A. Odetti, G. Bruzzone, M. Bibuli, E. Mandanici, L. Vittuari, P. Castaldi, M. Caccia et al., Underwater drone architecture for marine digital twin: lessons learned from sushi drop project. Sensors **22**(3), 744 (2022)
130. Ø.R. Smogeli, K.B. Ludvigsen, L. Jamt, B. Vik, H. Nordahl, L.T. Kyllingstad, K.K. Yum, H. Zhang, Open simulation platform-an open-source project for maritime system co-simulation, in *19th International Conference on Computer and IT Applications in the Maritime Industries* (Technische Universität Hamburg Harburg, 2020)
131. C. Gomes, C. Thule, D. Broman, P.G. Larsen, H. Vangheluwe, Co-simulation: a survey. ACM Comput. Surv. (CSUR) **51**(3), 1–33 (2018)
132. E. Durling, E. Palmkvist, M. Henningsson, FMI and IP protection of models: a survey of use cases and support in the standard, in *Proceedings of the 12th International Modelica Conference, Prague, Czech Republic, 15–17 May 2017*, no. 132 (Linköping University Electronic Press, 2017), pp. 329–335
133. B. Mishra, Applying digital twin in shipping and maritime (2020). https://seanews.co.uk/features/applying-digital-twin-in-shipping-and-maritime/. Date accessed 12 Feb 2024
134. S. Skjong, E. Pedersen, On the numerical stability in dynamical distributed simulations. Math. Comput. Simul. **163**, 183–203 (2019)
135. J. Bastian, C. Clauß, S. Wolf, P. Schneider, Master for co-simulation using FMI, in *Proceedings of the 8th International Modelica Conference; March 20th-22nd; Technical Univeristy; Dresden; Germany*, no. 63 (Linköping University Electronic Press, 2011), pp. 115–120
136. Ø. Smogeli, L. Bruun, L. Jamt, B. Vik, H. Nordahl, L. Kyllingstad, K. Yum, H. Zhang, Open simulation platform—an open-source project for maritime system co-simulation, in *19th International Conference on Computer and IT Applications in the Maritime Industries* (2020), pp. 239–253
137. G. Schweiger, C. Gomes, G. Engel, I. Hafner, J. Schoeggl, A. Posch, T. Nouidui, An empirical survey on co-simulation: promising standards, challenges and research needs. Simul. Model. Pract. Theory **95**, 148–163 (2019)
138. B. Schleich, N. Anwer, L. Mathieu, S. Wartzack, Shaping the digital twin for design and production engineering. CIRP Ann. Manuf. Technol. **66**(1), 141–144 (2017)
139. F. Perabo, D. Park, M. K. Zadeh, Ø. Smogeli, L. Jamt, Digital twin modelling of ship power and propulsion systems: application of the open simulation platform (OSP), in *2020 IEEE 29th International Symposium on Industrial Electronics (ISIE)* (IEEE, 2020), pp. 1265–1270

140. C. Dad, S. Vialle, M. Caujolle, J.-P. Tavella, M. Ianotto, Parallelization, distribution and scaling of multi-simulations on multi-core clusters, with Daccosim environment (2016)
141. R.M. Fujimoto, Parallel and distributed simulation systems, in *Proceeding of the 2001 Winter Simulation Conference (Cat. No. 01CH37304)*, vol. 1 (IEEE, 2001), pp. 147–157
142. A. Becue, E. Maia, L. Feeken, P. Borchers, I. Praca, A new concept of digital twin supporting optimization and resilience of factories of the future. Appl. Sci. **10**(13), 4482 (2020)
143. A. Junghanns, C. Gomes, C. Schulze, K. Schuch, R. Pierre, M. Blaesken, I. Zacharias, A. Pillekeit, K. Wernersson, T. Sommer, et al., The functional mock-up interface 3.0-new features enabling new applications, in *Modelica Conferences* (2021), pp. 17–26
144. M. Krammer, M. Benedikt, T. Blochwitz, K. Alekeish, N. Amringer, C. Kater, S. Materne, R. Ruvalcaba, K. Schuch, J. Zehetner, et al., The distributed co-simulation protocol for the integration of real-time systems and simulation environments, in *Proceedings of the 50th Computer Simulation Conference* (2018), pp. 1–14
145. M. Allmaras, A.B. Halblaub, H. Held, T. Schenk, A portable and secure package format for executable simulation modules based on web assembly, in *Modelica Conferences* (2021), pp. 443–452
146. E. Tantik, R. Anderl, Integrated data model and structure for the asset administration shell in Industrie 4.0. Procedia CIRP **60**, 86–91 (2017)
147. J. Köhler, H.-M. Heinkel, P. Mai, J. Krasser, M. Deppe, M. Nagasawa, Modelica-association-project "system structure and parameterization"–early insights (2016)
148. Open Simulation Platform, Osp interface specification (2022). https://opensimulationplatform.com/specification/. Date accessed 12 Feb 2024)
149. C.M. Legaard, D. Tola, T. Schranz, H.D. Macedo, P.G. Larsen, A universal mechanism for implementing functional mock-up units, in *SIMULTECH* (2021), pp. 121–129
150. L.I. Hatledal, E. Fagerhaug, Enhancing SSP creation using SSPGEN," in *Modelica Conferences* (2022), pp. 115–119
151. S. Sadjina, L.T. Kyllingstad, S. Skjong, E. Pedersen, Energy conservation and power bonds in co-simulations: non-iterative adaptive step size control and error estimation. Eng. Comput. **33**, 607–620 (2017)
152. S. Sadjina, L.T. Kyllingstad, M. Rindarøy, S. Skjong, V. Æsøy, E. Pedersen, Distributed co-simulation of maritime systems and operations. J. Offshore Mech. Arct. Eng. **141**(1) (2019)
153. C. Gomes, C. Thule, D. Broman, P.G. Larsen, H. Vangheluwe, Co-simulation: state of the art (2017). arXiv:1702.00686
154. L.I. Hatledal, A. Styve, G. Hovland, H. Zhang, A language and platform independent co-simulation framework based on the functional mock-up interface. IEEE Access **7**, 109328–109339 (2019)
155. G. Li, F.T. Holmeset, H. Zhang, Toward remote control center for marine operation: a case study of R/V Gunnerus, in *2023 11th International Conference on Control, Mechatronics and Automation (ICCMA)* (IEEE, 2023), pp. 183–188
156. R. Skulstad, G. Li, T.I. Fossen, B. Vik, H. Zhang, A hybrid approach to motion prediction for ship docking—integration of a neural network model into the ship dynamic model. IEEE Trans. Instrum. Meas. **70** (2021)
157. G. Li, H. Zhang, B. Kawan, H. Wang, O.L. Osen, A. Styve, Analysis and modeling of sensor data for ship motion prediction, in *OCEANS 2016-Shanghai* (IEEE, 2016), pp. 1–7
158. Y. Meng, X. Zhang, J. Zhu, Parameter identification of ship motion mathematical model based on full-scale trial data. Int. J. Nav. Arch. Ocean. Eng. **14**, 100437 (2022)
159. T. Wang, G. Li, L.I. Hatledal, R. Skulstad, V. Æsøy, H. Zhang, Incorporating approximate dynamics into data-driven calibrator: a representative model for ship maneuvering prediction. IEEE Trans. Ind. Inf. **18**(3), 1781–1789 (2021)
160. M. Kanazawa, R. Skulstad, T. Wang, G. Li, L.I. Hatledal, H. Zhang, A physics-data co-operative ship dynamic model for a docking operation. IEEE Sens. J. **22**(11), 11173–11183 (2022)
161. D.C. Karnopp, D.L. Margolis, R.C. Rosenberg, *System Dynamics: Modeling, Simulation, and Control of Mechatronic Systems*, 5th edn. (New Jersey, Wiley, Hoboken, 2012), pp.37–48

162. W. Borutzky, Bond graph modeling from an object oriented modeling point of view. Simul. Pract. Theory **7**(5), 439–461 (1999)
163. J.F. Broenink, Bond-graph modeling in Modelica, in *Proceedings of the 9th European Simulation Symposium* (Passau Germany, 1997), pp. 19–22
164. F.E. Cellier, À. Nebot, The modelica bond graph library, in *Proceedings of the 4th International Modelica Conference, Hamburg, Germany* (2005), pp. 57–65
165. J.F. Broenink, Object-oriented modeling with bond graphs and Modelica. Simul. Ser. **31**, 163–168 (1999)
166. T. Pulecchi, F. Casella, M. Lovera, Object-oriented modelling for spacecraft dynamics: tools and applications. Simul. Model. Pract. Theory **18**(1), 63–86 (2010)
167. A. Ross, Nonlinear manoeuvring models for ships: a Lagrangian approach (2008)
168. *Mechanical Systems with Nonlinear Geometry*, Chap. 9 (Wiley, 2012), pp. 411–469
169. M. Christie, A. Cliffe, P. Dawid, S.S. Senn, *Simplicity, Complexity and Modelling*, vol. 114 (Wiley, 2011)
170. D. Huang, D. Zhao, L. Wei, Z. Wang, Y. Du, Modeling and analysis in marine big data: advances and challenges. Math. Probl. Eng. **2015** (2015)
171. C. Fan, M. Chen, X. Wang, J. Wang, B. Huang, A review on data preprocessing techniques toward efficient and reliable knowledge discovery from building operational data. Front. Energy Res. **9**, 652801 (2021)
172. Z. Cui, Y. Cao, G. Wu, H. Liu, Z. Qiu, C. Chen, Research on preprocessing technology of building energy consumption monitoring data based on machine learning algorithm. Build. Sci. **34**(2), 94–99 (2018)
173. P. Kang, Locally linear reconstruction based missing value imputation for supervised learning. Neurocomputing **118**, 65–78 (2013)
174. C. Fan, F. Xiao, H. Madsen, D. Wang, Temporal knowledge discovery in big bas data for building energy management. Energy Build. **109**, 75–89 (2015)
175. Z.J. Yu, F. Haghighat, B.C. Fung, L. Zhou, A novel methodology for knowledge discovery through mining associations between building operational data. Energy Build. **47**, 430–440 (2012)
176. R. Preece, J.V. Milanović, Efficient estimation of the probability of small-disturbance instability of large uncertain power systems. IEEE Trans. Power Syst. **31**(2), 1063–1072 (2015)
177. V. Kumar, S. Minz, Feature selection: a literature review. SmartCR **4**(3), 211–229 (2014)
178. X. Yu, S. Ergan, G. Dedemen, A data-driven approach to extract operational signatures of HVAC systems and analyze impact on electricity consumption. Appl. Energy **253**, 113497 (2019)
179. S. Wang, X. Yao, Multiclass imbalance problems: analysis and potential solutions. IEEE Trans. Syst., Man, Cybern. Part B (Cybernetics) **42**(4), 1119–1130 (2012)
180. P. Branco, L. Torgo, R.P. Ribeiro, A survey of predictive modeling on imbalanced domains. ACM Comput. Surv. (CSUR) **49**(2), 1–50 (2016)
181. J. Van Hulse, T.M. Khoshgoftaar, A. Napolitano, Experimental perspectives on learning from imbalanced data, in *Proceedings of the 24th International Conference on Machine Learning* (2007), pp. 935–942
182. A. Saltelli, S. Tarantola, F. Campolongo, M. Ratto, et al., *Sensitivity Analysis in Practice: A Guide to Assessing Scientific Models*, vol. 1 (2004). Wiley Online Library
183. S. Razavi, A. Jakeman, A. Saltelli, C. Prieur, B. Iooss, E. Borgonovo, E. Plischke, S.L. Piano, T. Iwanaga, W. Becker et al., The future of sensitivity analysis: an essential discipline for systems modeling and policy support. Environ. Model. Softw. **137**, 104954 (2021)
184. F. Pianosi, K. Beven, J. Freer, J.W. Hall, J. Rougier, D.B. Stephenson, T. Wagener, Sensitivity analysis of environmental models: a systematic review with practical workflow. Environ. Model. Softw. **79**, 214–232 (2016)
185. M.-J. Shin, J.H. Guillaume, B.F. Croke, A.J. Jakeman, Addressing ten questions about conceptual rainfall-runoff models with global sensitivity analyses in R. J. Hydrol. **503**, 135–152 (2013)

186. M. Cao, L. Pan, Y. Gao, D. Novák, Z. Ding, D. Lehký, X. Li, Neural network ensemble-based parameter sensitivity analysis in civil engineering systems. Neural Comput. Appl. **28**, 1583–1590 (2017)
187. J. Wang, H. Zhang, K. Zhang, N.R. Pal, Bilateral sensitivity analysis for understandable neural networks and its application to reservoir engineering (2020)
188. H. Hu, T. Lin, S. Wang, L.F. Rodriguez, A cybergis approach to uncertainty and sensitivity analysis in biomass supply chain optimization. Appl. Energy **203**, 26–40 (2017)
189. Y. Medina, E. Muñoz, Analysis of the relative importance of model parameters in watersheds with different hydrological regimes. Water **12**(9), 2376 (2020)
190. H.V. Gupta, S. Razavi, Revisiting the basis of sensitivity analysis for dynamical earth system models. Water Resour. Res. **54**(11), 8692–8717 (2018)
191. S. Tarantola, D. Gatelli, S. Kucherenko, W. Mauntz et al., Estimating the approximation error when fixing unessential factors in global sensitivity analysis. Reliab. Eng. Syst. Saf. **92**(7), 957–960 (2007)
192. J.H. Guillaume, J.D. Jakeman, S. Marsili-Libelli, M. Asher, P. Brunner, B. Croke, M.C. Hill, A.J. Jakeman, K.J. Keesman, S. Razavi et al., Introductory overview of identifiability analysis: a guide to evaluating whether you have the right type of data for your modeling purpose. Environ. Model. Softw. **119**, 418–432 (2019)
193. D. Partington, M.J. Knowling, C.T. Simmons, P.G. Cook, Y. Xie, T. Iwanaga, C. Bouchez, Worth of hydraulic and water chemistry observation data in terms of the reliability of surface water-groundwater exchange flux predictions under varied flow conditions. J. Hydrol. **590**, 125441 (2020)
194. S. Tarantola, N. Giglioli, J. Jesinghaus, A. Saltelli, Can global sensitivity analysis steer the implementation of models for environmental assessments and decision-making? Stoch. Env. Res. Risk Assess. **16**, 63–76 (2002)
195. A. Saltelli, K. Aleksankina, W. Becker, P. Fennell, F. Ferretti, N. Holst, S. Li, Q. Wu, Why so many published sensitivity analyses are false: a systematic review of sensitivity analysis practices. Environ. Model. Softw. **114**, 29–39 (2019)
196. A. Saltelli, M. Ratto, T. Andres, F. Campolongo, J. Cariboni, D. Gatelli, M. Saisana, S. Tarantola, *Global Sensitivity Analysis: The Primer* (Wiley, 2008)
197. R.A. Howard, Uncertainty about probability: a decision analysis perspective. Risk Anal. **8**(1), 91–98 (1988)
198. R.B. Gramacy, M. Taddy et al., *Categorical inputs, sensitivity analysis, optimization and importance tempering with TGP version 2, an R package for treed gaussian process models* (University of Cambridge Statistical Laboratory Technical Report, 2009)
199. S. Burhenne, D. Jacob, G.P. Henze, Uncertainty analysis in building simulation with Monte Carlo techniques. Proc. SIMBUILD **4**(1), 419–426 (2010)
200. B. Devenish, P. Francis, B. Johnson, R. Sparks, D. Thomson, Sensitivity analysis of dispersion modeling of volcanic ash from eyjafjallajökull in May 2010. J. Geophys. Res.: Atmos. **117**(D20) (2012)
201. M.D. Morris, Factorial sampling plans for preliminary computational experiments. Technometrics **33**(2), 161–174 (1991)
202. W. Tian, A review of sensitivity analysis methods in building energy analysis. Renew. Sustain. Energy Rev. **20**, 411–419 (2013)
203. F. Campolongo, J. Kleijnen, T. Andres, Screening methods, in Wiley Series in Probability & Statistics (2000)
204. O. Rakovec, M.C. Hill, M. Clark, A. Weerts, A. Teuling, R. Uijlenhoet, Distributed evaluation of local sensitivity analysis (DELSA), with application to hydrologic models. Water Resour. Res. **50**(1), 409–426 (2014)
205. P.C. Young, G.M. Hornberger, R.C. Spear, Modeling badly defined systems: some further thoughts, in *Proceedings SIMSIG Conference, Canberra* (1978), pp. 24–32
206. R.C. Spear, G. Hornberger, Eutrophication in peel inlet-ii. identification of critical uncertainties via generalized sensitivity analysis. Water Res. **14**(1), 43–49 (1980)

207. F. Pianosi, T. Wagener, A simple and efficient method for global sensitivity analysis based on cumulative distribution functions. Environ. Model. Softw. **67**, 1–11 (2015)
208. Y. Tang, P. Reed, T. Wagener, K. Van Werkhoven, Comparing sensitivity analysis methods to advance lumped watershed model identification and evaluation. Hydrol. Earth Syst. Sci. **11**(2), 793–817 (2007)
209. R. Pastres, K. Chan, C. Solidoro, C. Dejak, Global sensitivity analysis of a shallow-water 3d eutrophication model. Comput. Phys. Commun. **117**(1–2), 62–74 (1999)
210. F. Minunno, M. van Oijen, D. Cameron, J. Pereira, Selecting parameters for Bayesian calibration of a process-based model: a methodology based on canonical correlation analysis. SIAM/ASA J. Uncertain. Quantif. **1**(1), 370–385 (2013)
211. E.B. Harper, J.C. Stella, A.K. Fremier, Global sensitivity analysis for complex ecological models: a case study of riparian cottonwood population dynamics. Ecol. Appl. **21**(4), 1225–1240 (2011)
212. R. Singh, T. Wagener, R. Crane, M. Mann, L. Ning, A vulnerability driven approach to identify adverse climate and land use change combinations for critical hydrologic indicator thresholds: Application to a watershed in Pennsylvania, USA. Water Resour. Res. **50**(4), 3409–3427 (2014)
213. I.M. Sobol, Sensitivity analysis for non-linear mathematical models. Math. Model. Comput. Exp. **1**, 407–414 (1993)
214. E. Ryan, O. Wild, A. Voulgarakis, L. Lee, Fast sensitivity analysis methods for computationally expensive models with multi-dimensional output. Geosci. Model. Dev. **11**(8), 3131–3146 (2018)
215. A. Saltelli, S. Tarantola, K.-S. Chan, A quantitative model-independent method for global sensitivity analysis of model output. Technometrics **41**(1), 39–56 (1999)
216. W. Chen, R. Jin, A. Sudjianto, Analytical variance-based global sensitivity analysis in simulation-based design under uncertainty. J. Mech. Des. **127**(5), 875–886 (2005)
217. K. Cheng, Z. Lu, C. Ling, S. Zhou, Surrogate-assisted global sensitivity analysis: an overview. Struct. Multidiscip. Optim. **61**, 1187–1213 (2020)
218. X. Luo, Z. Lu, X. Xu, Reproducing kernel technique for high dimensional model representations (HDMR). Comput. Phys. Commun. **185**(12), 3099–3108 (2014)
219. I.M. Sobol', Theorems and examples on high dimensional model representation. Reliab. Eng. Syst. Saf. **79**(2), 187–193 (2003)
220. G. Li, S.-W. Wang, C. Rosenthal, H. Rabitz, High dimensional model representations generated from low dimensional data samples. I. mp-Cut-HDMR. J. Math. Chem. **30**, 1–30 (2001)
221. M. Ratto, A. Pagano, P. Young, State dependent parameter metamodelling and sensitivity analysis. Comput. Phys. Commun. **177**(11), 863–876 (2007)
222. S. Ulaganathan, I. Couckuyt, T. Dhaene, J. Degroote, E. Laermans, High dimensional kriging metamodelling utilising gradient information. Appl. Math. Model. **40**(9–10), 5256–5270 (2016)
223. I. Abdallah, C. Lataniotis, B. Sudret, Parametric hierarchical kriging for multi-fidelity aero-servo-elastic simulators-application to extreme loads on wind turbines. Probab. Eng. Mech. **55**, 67–77 (2019)
224. Z. Wu, D. Wang, Z. Jiang, W. Zhang et al., Unified estimate of gaussian kernel width for surrogate models. Neurocomputing **203**, 41–51 (2016)
225. F. Pianosi, F. Sarrazin, T. Wagener, A MATLAB toolbox for global sensitivity analysis. Environ. Model. Softw. **70**, 80–85 (2015)
226. E. Borgonovo, A new uncertainty importance measure. Reliab. Eng. Syst. Saf. **92**(6), 771–784 (2007)
227. W. Yun, Z. Lu, X. Jiang, An efficient method for moment-independent global sensitivity analysis by dimensional reduction technique and principle of maximum entropy. Reliab. Eng. Syst. Saf. **187**, 174–182 (2019)
228. F. Pianosi, T. Wagener, A simple and efficient method for global sensitivity analysis based on cumulative distribution functions. Environ. Model. Softw. **67**, 1–11 (2015)

229. S. Razavi, H.V. Gupta, A new framework for comprehensive, robust, and efficient global sensitivity analysis: 1. theory. Water Resour. Res. **52**(1), 423–439 (2016)
230. R. Sheikholeslami, S. Razavi, A fresh look at variography: measuring dependence and possible sensitivities across geophysical systems from any given data. Geophys. Res. Lett. **47**(20), e2020GL089829 (2020)
231. A. Saltelli, A short comment on statistical versus mathematical modelling. Nat. Commun. **10**(1), 3870 (2019)
232. P. Pires da Silva, S. Sutulo, C. Guedes Soares, Sensitivity analysis of ship manoeuvring mathematical models. J. Mar. Sci. Eng. **11**(2), 416 (2023)
233. S. Li, T. Wang, G. Li, H. Zhang, Ship maneuvering model optimization for improved identification with less excitation. Ocean Eng. **280**, 114540 (2023)
234. X. Cheng, S. Chen, C. Diao, M. Liu, G. Li, H. Zhang, Simplifying neural network based model for ship motion prediction: a comparative study of sensitivity analysis, in *International Conference on Offshore Mechanics and Arctic Engineering*, vol. 57632 (American Society of Mechanical Engineers, 2017), p. V001T01A016
235. C. Wang, T.S. Nord, G. Ziemer, G. Li, Towards uncertainty and sensitivity analysis for modal parameters identification during ice-structure interaction. Ocean Eng. **277**, 114224 (2023)
236. K. Maute, M. Nikbay, C. Farhat, Coupled analytical sensitivity analysis and optimization of three-dimensional nonlinear aeroelastic systems. AIAA J. **39**(11), 2051–2061 (2001)
237. M.A. Gammon, Optimization of fishing vessels using a multi-objective genetic algorithm. Ocean. Eng. **38**(10), 1054–1064 (2011)
238. Q. Liu, B. Feng, Z. Liu, H. Zhang, The improvement of a variance-based sensitivity analysis method and its application to a ship hull optimization model. J. Mar. Sci. Technol. **22**, 694–709 (2017)
239. K. Liu, X. Xin, J. Ma, J. Zhang, Q. Yu, Sensitivity analysis of ship traffic in restricted two-way waterways considering the impact of lNG carriers. Ocean Eng. **192**, 106556 (2019)
240. Maritime safety committee (MSC 103/21/add.1, annex 8) Regulatory scoping exercise for the use of maritime autonomous surface ships (MASS). Report of the working group (2021)
241. G. Li, D. Ouyang, L. Yang, Q. Li, K. Tian, B. Wu, G. Guo, Cross-subject EEG linear domain adaption based on batch normalization and depthwise convolutional neural network. Knowl.-Based Syst. **280**, 111011 (2023)
242. G. Li, R. Mao, H.P. Hildre, H. Zhang, Visual attention assessment for expert-in-the-loop training in a maritime operation simulator. IEEE Trans. Ind. Inf. **16**(1), 522–531 (2019)
243. B. Wu, L. Zhao, S.R.T. Sunilkumar, H.P. Hildre, H. Zhang, G. Li, Visual attention analysis for critical operations in maritime collision avoidance, in *2022 IEEE 17th International Conference on Control & Automation (ICCA)* (IEEE, 2022), pp. 285–2920
244. B. Wu, P. Han, M. Kanazawa, H.P. Hildre, L. Zhao, H. Zhang, G. Li, A camera-based deep-learning solution for visual attention zone recognition in maritime navigational operations, in *2022 IEEE/RSJ International Conference on Intelligent Robots and Systems (IROS)* (IEEE, 2022), pp. 699–706
245. T.I. Fossen, *Handbook of Marine Craft Hydrodynamics and Motion Control* (Wiley, 2011)
246. N. Smogeli, *Control of Marine Propellers From Normal to Extreme Conditions* (2006)
247. B. Kawan, H. Wang, G. Li, K. Chhantyal, Data-driven modeling of ship motion prediction based on support vector regression, in *Proceedings of the 58th SIMS Conference* (2017), pp. 350–354
248. G. Cademartori, L. Oneto, F. Valdenazzi, A. Coraddu, A. Gambino, D. Anguita, A review on ship motions and quiescent periods prediction models. Ocean Eng. **280**(2022), 114822 (2023)
249. M. Kanazawa, R. Skulstad, T. Wang, G. Li, L.I. Hatledal, H. Zhang, A physics-data co-operative ship dynamic model for a docking operation. IEEE Sens. J. **22**(11), 11173–11183 (2022)
250. S. Ben Taieb, A. Sorjamaa, G. Bontempi, Multiple-output modeling for multi-step-ahead time series forecasting. Neurocomputing **73**(10–12), 1950–1957 (2010)
251. H. Jiang, S.L. Duan, L. Huang, Y. Han, H. Yang, Q. Ma, Scale effects in AR model real-time ship motion prediction. Ocean Eng. **203**, 107202 (2020)

252. M. Kanazawa, R. Skulstad, G. Li, L.I. Hatledal, H. Zhang, A multiple-output hybrid ship trajectory predictor with consideration for future command assumption. IEEE Sens. J. **21**(23), 27124–27135 (2021)
253. S.B. Taieb, A.F. Atiya, A bias and variance analysis for multistep-ahead time series forecasting. IEEE Trans. Neural Netw. Learn. Syst. **27**(1), 62–76 (2016)
254. R. Skjetne, Ø.N. Smogeli, T.I. Fossen, A nonlinear ship manoeuvering modelâŁ¯: identification and adaptive control with experiments for a model ship. MIC—model. Identif. Control **25**(1), 3–27 (2004)
255. A. Ross, V. Hassani, Ø. Selvik, E. Ringen, D. Fathi, Identification of nonlinear manoeuvring models for marine vessels using planar motion mechanism test, in *Proceedings of the ASME 2015 34th International Conference on Ocean, Offshore and Arctic Engineering* (2015), pp. 1–7
256. M.A. Roehrl, T.A. Runkler, V. Brandtstetter, M. Tokic, S. Obermayer, Modeling system dynamics with physics-informed neural networks based on Lagrangian mechanics. IFAC-PapersOnLine **53**, 9195–9200 (Elsevier Ltd., 2020)
257. R. Skulstad, *Data-Based Modelling of Ships*. Ph.D. thesis, Norwegian University of Science and Technology (2021)
258. R. Skulstad, G. Li, T.I. Fossen, B. Vik, H. Zhang, Dead reckoning of dynamically positioned ships: using an efficient recurrent neural network. IEEE Robot. Autom. Mag. **26**(3), 39–51 (2019)
259. T.I. Fossen, *Handbook of Marine Craft Hydrodynamics and Motion Control*, 2nd edn. (Wiley, 2021)
260. R.H. Rogne, T.H. Bryne, T.I. Fossen, T.A. Johansen, MEMS-based inertial navigation on dynamically positioned ships: dead reckoning. IFAC-PapersOnLine **49**(23), 139–146 (2016)
261. K.E. Knutsen, G. Manno, B. Vartdal, Beyond condition monitoring in the maritime industry. DNV GL Strategic Research & Inovation Position Paper (2014)
262. A.L. Ellefsen, V. Æsøy, S. Ushakov, H. Zhang, A comprehensive survey of prognostics and health management based on deep learning for autonomous ships. IEEE Trans. Reliab. **68**(2), 720–740 (2019)
263. P.L. Dussault, Creating a closed loop environment for condition-based maintenance plus (CMB+) and prognostics health management," in *2007 IEEE Autotestcon* (2007), pp. 327–331
264. F. Camci, G.S. Valentine, K. Navarra, Methodologies for integration of PHM systems with maintenance data, in *2007 IEEE Aerospace Conference* (2007), pp. 1–9
265. K.M. Janasak, R.R. Beshears, Diagnostics to prognostics—a product availability technology evolution, in *2007 Annual Reliability and Maintainability Symposium* (2007), pp. 113–118
266. J. Lee, F. Wu, W. Zhao, M. Ghaffari, L. Liao, D. Siegel, Prognostics and health management design for rotary machinery systems-reviews, methodology and applications. Mech. Syst. Signal Process. **42**(1), 314–334 (2014)
267. W. Elghazel, J. Bahi, C. Guyeux, M. Hakem, K. Medjaher, N. Zerhouni, Dependability of wireless sensor networks for industrial prognostics and health management. Comput. Ind. **68**, 1–15 (2015)
268. A.K. Jardine, D. Lin, D. Banjevic, A review on machinery diagnostics and prognostics implementing condition-based maintenance. Mech. Syst. Signal Process. **20**(7), 1483–1510 (2006)
269. P.W. Kalgren, C.S. Byington, M.J. Roemer, M.J. Watson, Defining PHM, a lexical evolution of maintenance and logistics, in *2006 IEEE Autotestcon* (IEEE, 2016), pp. 353–358
270. G. Vachtsevanos, F. Lewis, M. Roemer, A. Hess, B. Wu, Fault prognosis, in *Intelligent Fault Diagnosis and Prognosis for Engineering Systems* (2007), pp. 284–354
271. D.A. Tobon-Mejia, K. Medjaher, N. Zerhouni, G. Tripot, A data-driven failure prognostics method based on mixture of gaussians hidden Markov models. IEEE Trans. Reliab. **61**(2), 491–503 (2012)
272. J.Z. Sikorska, M. Hodkiewicz, L. Ma, Prognostic modelling options for remaining useful life estimation by industry. Mech. Syst. Signal Process. **25**(5), 1803–1836 (2011)

273. D. McDonnell, N. Balfe, S. Al-Dahidi, G.E. O'Donnell, Designing for human-centred decision support systems in PHM, in *PHM Society European Conference*, vol. 2 (2014)
274. M. Yasar, T.E. Lovett, PHM decision support under uncertainty, in *Annual Conference of the PHM Society*, vol. 8 (2016)
275. N. Iyer, K. Goebel, and P. Bonissone, Framework for post-prognostic decision support, in *2006 IEEE Aerospace Conference* (IEEE, 2006), p. 10
276. D.J. Power, Web-based and model-driven decision support systems: concepts and issues, in *AMCIS 2000 Proceedings* (2000), p. 387
277. S. Yin, S.X. Ding, X. Xie, H. Luo, A review on basic data-driven approaches for industrial process monitoring. IEEE Trans. Ind. Electron. **61**(11), 6418–6428 (2014)
278. P. Baraldi, F. Di Maio, D. Genini, E. Zio, Comparison of data-driven reconstruction methods for fault detection. IEEE Trans. Reliab. **64**(3), 852–860 (2015)
279. D. Yu, L. Deng, Deep learning and its applications to signal and information processing [exploratory DSP]. IEEE Signal Process. Mag. **28**(1), 145–154 (2010)
280. J. Qiu, Q. Wu, G. Ding, Y. Xu, S. Feng, A survey of machine learning for big data processing. EURASIP J. Adv. Signal Process. **2016**, 1–16 (2016)
281. W. Liu, Z. Wang, X. Liu, N. Zeng, Y. Liu, F.E. Alsaadi, A survey of deep neural network architectures and their applications. Neurocomputing **234**, 11–26 (2017)
282. Z.-H. Ling, S.-Y. Kang, H. Zen, A. Senior, M. Schuster, X.-J. Qian, H.M. Meng, L. Deng, Deep learning for acoustic modeling in parametric speech generation: a systematic review of existing techniques and future trends. IEEE Signal Process. Mag. **32**(3), 35–52 (2015)
283. K. Noda, Y. Yamaguchi, K. Nakadai, H.G. Okuno, T. Ogata, Audio-visual speech recognition using deep learning. Appl. Intell. **42**, 722–737 (2015)
284. A. Krizhevsky, I. Sutskever, G.E. Hinton, Imagenet classification with deep convolutional neural networks, in *Advances in Neural Information Processing Systems*, vol. 25 (2012)
285. I. Sutskever, O. Vinyals, Q.V. Le, Sequence to sequence learning with neural networks, in *Advances in Neural Information Processing Systems*, vol. 27 (2014)
286. M. Längkvist, L. Karlsson, A. Loutfi, A review of unsupervised feature learning and deep learning for time-series modeling. Pattern Recognit. Lett. **42**, 11–24 (2014)
287. O. Levander, Autonomous ships on the high seas. IEEE Spectr. **54**(2), 26–31 (2017)
288. R. Jalonen, R. Tuominen, M. Wahlström, Remote and autonomous ships-the next steps: safety and security, in *Rolls-Royce, Buckingham Gate, London: The Advanced Autonomous Waterborne Applications (AAWA)* (2016), pp. 56–73
289. B.-M. Batalden, P. Leikanger, P. Wide, Towards autonomous maritime operations, in *2017 IEEE International Conference on Computational Intelligence and Virtual Environments for Measurement Systems and Applications (CIVEMSA)* (IEEE, 2017), pp. 1–6
290. S. Salvador, P. Chan, Learning states and rules for detecting anomalies in time series. Appl. Intell. **23**, 241–255 (2005)
291. P. Sun, S. Chawla, B. Arunasalam, Mining for outliers in sequential databases, in *Proceedings of the 2006 SIAM International Conference on Data Mining* (SIAM, 2006), pp. 94–105
292. J. An, S. Cho, Variational autoencoder based anomaly detection using reconstruction probability. Spec. Lect. IE **2**(1), 1–18 (2015)
293. H. Zhao, Y. Wang, J. Duan, C. Huang, D. Cao, Y. Tong, B. Xu, J. Bai, J. Tong, Q. Zhang, Multivariate time-series anomaly detection via graph attention network, in *2020 IEEE International Conference on Data Mining (ICDM)* (IEEE, 2020), pp. 841–850
294. D. Park, Y. Hoshi, C.C. Kemp, A multimodal anomaly detector for robot-assisted feeding using an LSTM-based variational autoencoder. IEEE Robot. Autom. Lett. **3**(3), 1544–1551 (2018)
295. T. Kurtoglu, O.J. Mengshoel, S. Poll, A framework for systematic benchmarking of monitoring and diagnostic systems, in *2008 International Conference on Prognostics and Health Management* (IEEE, 2008), pp. 1–13
296. A.L. Ellefsen, P. Han, X. Cheng, F.T. Holmeset, V. Æsøy, H. Zhang, Online fault detection in autonomous ferries: using fault-type independent spectral anomaly detection. IEEE Trans. Instrum. Meas. **69**(10), 8216–8225 (2020)

297. F.T. Liu, K.M. Ting, Z.-H. Zhou, Isolation forest, in *2008 Eighth IEEE International Conference on Data Mining* (IEEE, 2008), pp. 413–422
298. P. Han, G. Li, R. Skulstad, S. Skjong, H. Zhang, A deep learning approach to detect and isolate thruster failures for dynamically positioned vessels using motion data. IEEE Trans. Instrum. Meas. **70**, 1–11 (2020)
299. D. Kwon, M.H. Azarian, M. Pecht, Prognostics of interconnect degradation using RF impedance monitoring and sequential probability ratio test. Int. J. Perform. Eng. **6**(5), 443 (2010)
300. A. Saxena, K. Goebel, D. Simon, N. Eklund, Damage propagation modeling for aircraft engine run-to-failure simulation, in *2008 International Conference on Prognostics and Health Management* (IEEE, 2008), pp. 1–9
301. A.L. Ellefsen, S. Ushakov, V. Æsøy, H. Zhang, Validation of data-driven labeling approaches using a novel deep network structure for remaining useful life predictions. IEEE Access **7**, 71563–71575 (2019)
302. L. Jayasinghe, T. Samarasinghe, C. Yuen, J.C.N. Low, S.S. Ge, Temporal convolutional memory networks for remaining useful life estimation of industrial machinery (2018). arXiv:1810.05644
303. N. Srivastava, G. Hinton, A. Krizhevsky, I. Sutskever, R. Salakhutdinov, Dropout: a simple way to prevent neural networks from overfitting. J. Mach. Learn. Res. **15**, 1929–1958 (2014)
304. X. Li, Q. Ding, J.-Q. Sun, Remaining useful life estimation in prognostics using deep convolution neural networks. Reliab. Eng. Syst. Saf. **172**, 1–11 (2018)
305. S. Zheng, K. Ristovski, A. Farahat, C. Gupta, Long short-term memory network for remaining useful life estimation, in *2017 IEEE International Conference on Prognostics and Health Management (ICPHM)* (IEEE, 2017), pp. 88–95
306. A.L. Ellefsen, E. Bjørlykhaug, V. Æsøy, S. Ushakov, H. Zhang, Remaining useful life predictions for turbofan engine degradation using semi-supervised deep architecture. Reliab. Eng. Syst. Saf. **183**, 240–251 (2019)
307. D.J. Power, R. Sharda, Decision support systems, in *Springer Handbook of Automation*. ed. by S.Y. Nof, Chap. 87, (Springer, Berlin Heidelberg, 2009), pp.1539–1548
308. N. Iyer, K. Goebel, P. Bonissone, Framework for post-prognostic decision support, in *2006 IEEE Aerospace Conference* (IEEE, 2006), p. 10
309. D. McDonnell, N. Balfe, S. Al-Dahidi, G. O'Donnell, Designing for human-centred decision support systems in PHM, in *European Conference of the Prognostics and Health Management Society* (IEEE, 2014)
310. H. Wang, G. Li, B. Kawan, H. Zhang, Neural-network-based modelling and analysis for time series prediction of ship motion. Ship Technol. Res. **64**(1), 30–39 (2017)
311. *Deterministic Predictions of Vessel Responses Based on Past Measurements*, vol. All Days of *International Ocean and Polar Engineering Conference* (2017)
312. P. Major, H. Zhang, H.P. Hildre, M. Edet, Virtual prototyping of offshore operations: a review. Ship Technol. Res. **68**(2), 84–101 (2021)
313. A. Ciuriuc, J.I. Rapha, R. Guanche, J.L. Domínguez-García, Digital tools for floating offshore wind turbines (FOWT): a state of the art. Energy Rep. **8**, 1207–1228 (2022)
314. SIMA (Last accessed 19 Feb 2023). https://www.dnvgl.com/services/marine-operations-and-mooring-analysis-software-sima-2324
315. B. Ataei, S. Yuan, Z. Ren, K.H. Halse, Effects of structural flexibility on the dynamic responses of low-height lifting mechanism for offshore wind turbine installation. Mar. Struct. **89**, 103399 (2023)
316. Z. Ren, Z. Jiang, R. Skjetne, Z. Gao, Development and application of a simulator for offshore wind turbine blades installation. Ocean Eng. **166**, 380–395 (2018)
317. *Mooring System Design and Analysis for a Floating Offshore Wind Turbine in Pantelleria*, vol. 11: Wind Energy of *Turbo Expo: Power for Land, Sea, and Air* (2022)
318. T. Adedipe, M. Shafiee, E. Zio, Bayesian network modelling for the wind energy industry: an overview. Reliab. Eng. Syst. Saf. **202**, 107053 (2020)

319. M.-Y. Cheng, Y.-F. Wu, Y.-W. Wu, S. Ndure, Fuzzy Bayesian schedule risk network for offshore wind turbine installation. Ocean Eng. **188**, 106238 (2019)
320. Y. Hadjoudj, R. Pandit, A review on data-centric decision tools for offshore wind operation and maintenance activities: challenges and opportunities. Energy Sci. Eng. **11**(4), 1501–1515 (2023)
321. W.J. North, C.H. Woodling, What made Apollo a success? Chap. 4 Apollo crew procedures, simulation, and flight planning. Astronaut. Aeronaut. 31–40 (1971)
322. J.I. Håvold, S. Nistad, A. Skiri, A. Ødegård, The human factor and simulator training for offshore anchor handling operators. Saf. Sci. **75**, 136–145 (2015)
323. A. Gepperth, B. Hammer, Incremental learning algorithms and applications, in *European Symposium on Artificial Neural Networks (ESANN)* (2016)
324. G.I. Parisi, R. Kemker, J.L. Part, C. Kanan, S. Wermter, Continual lifelong learning with neural networks: a review. Neural Netw. **113**, 54–71 (2019)
325. J. Kirkpatrick, R. Pascanu, N. Rabinowitz, J. Veness, G. Desjardins, A.A. Rusu, K. Milan, J. Quan, T. Ramalho, A. Grabska-Barwinska et al., Overcoming catastrophic forgetting in neural networks. Proc. Natl. Acad. Sci. **114**(13), 3521–3526 (2017)

Printed by Printforce, the Netherlands